THE END OF FOSSIL FUEL INSANITY

Clearing the Air
Before Cleaning the Air

Terry Etam

FriesenPress

Suite 300 - 990 Fort St
Victoria, BC, V8V 3K2
Canada

www.friesenpress.com

Copyright © 2019 by Terry Etam
First Edition — 2019

Cover design by Chris Beaudin (www.cbeaudin.com)

All rights reserved.

No part of this publication may be reproduced in any form, or by any means, electronic or mechanical, including photocopying, recording, or any information browsing, storage, or retrieval system, without permission in writing from FriesenPress.

ISBN
978-1-5255-4024-0 (Hardcover)
978-1-5255-4025-7 (Paperback)
978-1-5255-4026-4 (eBook)

1. SCIENCE, GLOBAL WARMING & CLIMATE CHANGE

Distributed to the trade by The Ingram Book Company

Table of Contents

Introduction xi

1. Energy as we (don't) know it 1
2. How did we get in this mess? 9
   PART I: Systems developed around fossil fuels
   LOST IN THE GOOD LIFE, WE HAVE NO RECOLLECTION OF HOW WE GOT HERE 9
   VERY BIG SYSTEMS DEVELOP WEIRDLY AND (NEARLY) PERMANENTLY 10
   MODIFYING OR REPLACING SYSTEMS IS UNBELIEVABLY DIFFICULT 13
   LAST BUT NOT LEAST: THE HUMAN SYSTEMS 17

3. How did we get in this mess? 19
   PART II: Overreliance on fossil fuels

4. How did we get in this mess? 23
   PART III: Divergence of opinions

5. Environmentalists are right
   – oil usage trajectory is unsustainable 29
   ROAD TRIP! ROAD TRIP! 29
   WHY IS THIS A PROBLEM FOR BOTH SIDES OF THE FOSSIL FUEL DEBATE? 31
   WAIT A MINUTE… HOW MANY CARS ARE WE TALKING ABOUT? 32
   WHAT ABOUT ELECTRIC VEHICLES? 34
   NO ONE IS GOING TO BE HAPPY 38
   THE ABSOLUTE VERSUS PER CAPITA DEBATE 39

6. The electrical utility guy is also right though.
   Wait, who's that again? 41
   DANGER, HIGH VOLTAGE – THIS ISN'T A SEWER PIPE YOU'RE MESSING WITH 43
   WORLD'S WORST ELCTRICAL GRID OVERVIEW, BUT IT'LL DO 43
   THE ELECTRICAL GRID 101 44
   IEC 61850 EXAMPLE – A CRITICAL, UNKNOWN STANDARD THAT SMART-GRIDS REQUIRE . . 47
   BEWARE OF LOONY EV PREDICTIONS 49
   MEGA BATTERIES COULD CHANGE EVERYTHING
   —BUT WE'VE BEEN SAYING THAT FOR A HUNDRED YEARS 50

7. Legacy of big oil 55
A FEW CENTURIES OF COLONIAL HABITS KICKSTART THE GLOBAL OIL PHENOMENON . . . 55
BIG OIL 56
WHAT WAS SO BAD ABOUT BIG OIL? 57
YOU GOT OIL? WANNA MAKE A DEAL? 58
CAN'T GO BACK 61
THE MARK ON PEOPLE 62
THE MARK ON GEOPOLITICS 64
THEY WALK AMONG US. ACTUALLY, THEY ARE US 67

8. What about coal, the "other" fossil fuel? 69
IT'S DISGUSTING, BUT EVERYONE'S DOING IT 70
THE ROLE OF COAL 71
HOW COME COAL GETS A FREE RIDE WHEN IT'S SO DISGUSTING? . . . 72

9. Rise of climate change as a global issue 77
FORTY YEARS AGO, THERE WAS TALK ABOUT ANOTHER ICE AGE, RIGHT? 77
WE'RE CRAPPING ON GLOBAL WARMING THEORY THEN, IT SOUNDS LIKE 80
CLIMATE CHANGE DENIERS 82
SHOW SOME RESPECT 83

10. Modern protesting: amateur hour no longer 85
PROTESTS THROUGH THE AGES—FROM THE NECESSARY TO THE NORMATIVE 85
THE SIXTIES OPENED THE MIND AND SHARPENED THE PROTESTING TOOLS 87
PROTESTING AND SOCIAL MEDIA 89

11. Out of touch: 1950^s PR strategies run head first into social media guerilla warfare 93
BIG, COMFORTABLE OIL 93
LOBBY? WHAT LOBBY? JUST DO WHAT WE TELL YOU OR TEN THOUSAND JOBS DISAPPEAR . 94
WHAT'S A FACE BOOK? NEVERMIND, WE'VE GOT A BUSINESS TO RUN 97
CLIMATE CHANGE BOMBSHELL 98

12. Key messaging warfare 101
LEARNING THE MESSAGING TRADE FROM TURTLES 101
EVEN THE GOOD STUFF THEY GET WRONG 105
THE EVIL NECESSITY THAT IS KEY MESSAGING 107
SO WHY DON'T THEY CHANGE? IT'S THE SYSTEMS, AGAIN 108
PR HAS ALWAYS BEEN CREEPY. NOW IT'S REALLY F___ING CREEPY.
   HERE'S WHAT THE PROS ARE UP TO 110
THE NEW BLACK OPS 111
EVERYTHIING OLD IS NEW AGAIN 112
CULTURE JAMMING PIONEERS WILL RUE THE DAY THEY EVER STARTED THIS 113

13. Lying with statistics 115
FROM THE HORSE'S MOUTH. 115
TACTICAL GENIUS – THE PLANTING OF THE 'OIL SANDS CARBON BOMB' MYTHICAL TREE . . 118
WHAT ABOUT THE OTHER SIDE?. 122
HALF-TRUTHS HAVE MANY FORMS 123

14. Not lying with statistics. 127
TESTING, TESTING 128
GLOBAL OIL PRODUCTION AND CONSUMPTION 129
RENEWABLE ENERGY PRODUCTION AND CONSUMPTION 131
WE GET IN BIG TROUBLE BY NOT UNDERSTANDING THESE CONCEPTS 132
AGAIN WITH THE SYSTEMS 133

15. Fears are real, but don't always make sense 135
GENERAL LIFE FEAR. 136
WHAT SHOULD SCARE US BUT DOESN'T 138
FEAR OF CLIMATE CHANGE 140
FEAR OF RUNNING OUT OF FOSSIL FUELS? 142

16. Consumption and population are both growing . . . 145
SOME THINGS WE CAN'T COMPREHEND DON'T MATTER, OTHERS MOST CERTAINLY DO. . . 146
THE WHOLE WORLD WANTS TO CONSUME LIKE WE DO TOO . . 148

17. The petroleum industry can do better 151
WANTED: CULTURAL REVOLUTION 151
STAKEHOLDER RELATIONS NEED TO BE MORE THAN A PLATITUDE 154
TIMES ARE CHANGING, EVEN IN THE OIL PATCH. 158
DESPERATELY NEEDED: A GROUND-UP GROWN-UP ATTEMPT AT OPENNESS,
  HONESTY, AND TRYING NOT TO SOUND FIFTY YEARS OUT OF DATE 160

18. Oil spills and other nastiness 163
SH_T DOES INDEED HAPPEN – INDUSTRIAL ACCIDENTS ARE AS HARD TO ERADICATE AS RATS . 164
THAT DOESN'T MEAN ALL ACCIDENTS ARE THE SAME 165
WAIT A MINUTE, IS THIS SOME SORT OF EXCULPATION OF PIPELINES AND ALL THEIR LEAKS? . 166
THREE EVENTS REDEFINED OIL TRANSPORTATION 167
EXXON VALDEZ TANKER SPILL 168
BP GULF OF MEXICO. 169
ENBRIDGE MICHIGAN 171
THE COSTS THEY ARE A CHANGIN' 173
BUT DO NOT THINK FOR A SECOND THAT ANYONE IS COMPLACENT ABOUT OIL SPILLS . . 174

19. Transporting energy – options and learnings . . . 175
TANKERS 177
PIPELINES 178

RAIL	179
TRUCK	180
POWER LINES	181
FINAL WORD	182

20. The futility of fighting pipelines or infrastructure – energy has to move ... 183
STOPPING THE FLOW OF A RIVER WITH A FISHING NET ... 183
GET OVER IT: LIMITING SUPPLY DOES NOT STOP CONSUMPTION ... 184
THE WORLD LOOKS A LOT DIFFERENT THAN FIFTY OR A HUNDRED YEARS AGO – A STUPID COMMENT BUT WE FORGET WHAT IT MEANS ... 185
YOUR ENERGY DOESN'T COME FROM WHERE IT USED TO, AND WE MUST DEAL WITH IT ... 187
THE BRIGHT SIDE OF A DISASTER – A SIMILAR SECOND ONE IS A LOT LESS LIKELY ... 189
YOU THINK YOU HATE OIL SPILLS? NOT HALF AS MUCH AS THE OTHER PIPELINE COMPANIES ... 190

21. But it never was about pipelines ... 193
ABUSING THE HELP ... 193
FIGHTING SYMPTOMS IS POINTLESS, BUT MAKES ONE HELL OF A GOOD STRATEGY ... 194

22. Habitat destruction vs. contamination vs. pollution vs. global warming ... 199
HABITAT DESTRUCTION ... 200
CONTAMINATION – THE RESULT OF DOING SOMETHING BADLY ... 202
POLLUTION - THE SUM EFFECT OF A CONSUMER'S LIFE ... 203
GLOBAL WARMING - AN ATTEMPT TO REDUCE CO2 EMISSIONS ... 204

23. Is Green Energy there yet? ... 207
GOING ALL GREEN – UNDERSTANDING WHERE WE'RE AT ... 211

24. A way forward, for those who want it ... 215
BRING ON THE FIVE DOLLAR DRINKING STRAWS ... 216
HYDROGEN ... 218
SWITCH FROM COAL TO NATURAL GAS ... 220
CARBON CAPTURE, ONE WAY OR ANOTHER ... 221
CARBON/CONSUMPTION TAXES OR REBATE SCHEMES ... 223
USE WHAT WE HAVE – BUILD A THOUSAND GREENHOUSES WITH OUR WASTE ENERGY ... 224

Epilogue ... 227

References ... 229

About the Author ... 235

"Out beyond ideas of right doing and wrongdoing, there is a field. I'll meet you there."

—Sufi poet Rumi

Pretty deep quote, wasn't that? Well, don't get your hopes up. I have no idea what Sufi means, and I don't know Rumi from a fence post (in fact, it's possible Sufi is the poet and I have that backwards). That should be embarrassing but oddly, I feel fine because the quote really works. Thanks Rumi. Or Sufi.

Introduction

Tomorrow morning, billions of people will get up, perform whatever hygienic processes that are necessary to remain in their respective social circles, and head out into a world of near-infinite possibilities. They will eat everything and anything imaginable. They will perform countless tasks, errands, and occupations to earn a living and keep their worlds moving. They will perform myriad other activities purely for amusement, some routine, some sporting, and some so bizarre they will leave God himself scratching his head.

Collectively, we'll do billions of things, from the sublime nothingness of laying on the couch watching TV to the thrill of extreme sports; from serving hamburgers to writing laws; from driving trucks to performing liposuction. We will do all that stuff because that's the infinite fabric of our way of life. Billions of people trying to sustain and amuse themselves will create a daily catalogue of productivity and motion and weirdness so huge and diverse that it can scarcely be imagined.

While we perform all these countless functions, hardly any of us will stop to wonder: How did all of this get here? I don't mean nature's contributions like birds and rocks and trees, but the stuff that makes up our daily existence. How did my food get here? How did my house? My toothbrush, my deodorant, my toast, my car, my fork? Are all those things as unfathomably permanent and "just there" as trees?

We'll leave the rocks and trees and birds to hard-core philosophers and post-grad unemployables to ponder. Instead, we will focus on the human-made world that we take for granted in a big way. Almost none of the participants in this global anthill of activity will stop to think about the underlying mechanics of what makes it all possible, and even those nerds that do won't cover ten percent of it.

No one does, and for any number of reasons. First, the entire infrastructure backbone just keeps working as long as everyone is doing their job, so we pay no attention and take it for granted. Second, it's too overwhelming to contemplate, even if we are interested. Third, it's really boring. Outside of engineers and other technical people, no one thinks about the pipes and wires and circuits and delivery mechanisms that get everything everywhere. No one talks about heating ducts or pipe couplings or valves. Why would you? But it all matters, every bit of it. Where did the metal that makes the utensils that feed us come from? Who assembled it all, and how? What about your cell phone? Those things are a chemical and metallurgical soup that sucks in elements that you never knew existed from the remotest regions of the earth, elements that have to be found and mined and processed and transported to China and assembled and then sent back to you. The same can be said for a hundred other things we encounter before we leave the front door, each with its own unique story and requirements for existence. We don't want to live without any of them.

* * *

If one day we're feeling philosophical and stop to ponder the world's most impressive human achievements, it's likely that forks won't make the top ten. Our minds tend to wander towards lists drawn up by globetrotting acquaintances or travel websites, or engineering feats like the pyramids of Egypt, a jet, or a laser. Or maybe we'd think of smaller, more personal things like a computer or cell phone, or other miraculous medical achievements like a doctor's ability to remove a tooth and implant a little bolt in your head on which to screw a new one. I have to admit, that's pretty cool, even if would rather saw my arm off than try it.

In the same way, it is unlikely that anyone's list would include our energy distribution system. An ungodly mess of pipes, refineries, tanker ships, and overhead power lines is not the stuff of tourist brochures or bucket-list must-sees.

This is a shame. Well, I suppose one can hardly call it a shame that you don't want to see the inside of a refinery before you die. Perhaps it's more

INTRODUCTION

of an injustice. Our energy distribution system keeps 7 billion people alive, and does so reliably. We've developed a staggeringly complex energy and food distribution system over the past century that has enabled a standard of living that was inconceivable two hundred years ago. It's become so reliable and uninteresting that it is part of the background noise for most people. The energy distribution system only gets attention when a dead fish or oil-covered bird appears, and then we loathe it.

To say that people don't appreciate these systems is not a "biting the hand that feeds" comparison, because that term is only appropriate if people understand the hand-food-bite relationship. With energy, they don't, and it's not even close.

Part of the reason they don't understand is that it's been a constant in almost everyone's lives for as long as they can remember, just like sewage systems have been—and we don't wax lyrical about those either. Things we grow up with, we take for granted. But the world is changing, and our energy diet is going to change also. We are going to be asked, or required, to rewire our energy distribution systems, and it is hard to map out what that will (or even should) look like when we don't understand what we have.

* * *

We are on the cusp of an energy revolution, one way or the other. Even for people that don't think about energy at all, the feeling is palpable. Everyone is excited by promises of a joyful bounty from wind and solar projects that sprout like spring flowers and promise freedom from what is perceived as the tyranny of control by either a bunch of unfriendly Middle East regimes or multinational oil corporations. This revolution hints at energy nirvana: nearly-free energy from an unlimited source, available whenever we want it, producible everywhere the wind blows or the sun shines. With it we can charge an electric vehicle (EV) and go wherever we want, and big oil can rot in hell.

Except it's not that way at all because of our massive, complex energy system—the same one that brought you a fork, powers the toaster, puts tires on your bike, fuels your car, and heats your house. We think that solar panels and wind turbines can replace the fossil fuel system. Some day they might, but the amount of work it's going to take, and the challenges we must overcome to make that happen, is a thousand miles the other side of formidable.

In ancient cities, modern amenities built up around the old streets, and effective transportation systems evolved. This was not necessarily efficient, but it was effective. A cow path is effective, but unless the lead cow was an engineer, it will wander this way and that. Perhaps a thousand years ago, the path became a road, which eventually had buildings on it. You can sit on such roads today in some European cities, oblivious to the bovine architect's contribution to the current state of affairs.

Imagine what it would take to make an old, historic city efficient in this day and age; that is, to replace all those crazy winding streets with a functional grid of right-angles. Any aspect of the project would be overwhelming. Imagine the approvals process or public consultation requirements, which, based on present observations, would last about four hundred years. Imagine coordinating utilities so existing businesses could continue to operate without stoppage. Imagine all the rewiring and re-piping and reconstruction that would be required. Now, imagine doing that for every city in the world.

That, in effect, is what switching from a hydrocarbon-based society to a renewable one will be like. It is a mistake of monumental proportions to think of it as plug and play, that renewable power sources like solar and wind farms can simply be patched into the system and it will all work. It works well when their contributions are at five percent, but when we ask them to contribute the majority of what we use, we find ourselves in uncharted waters. There are infrastructure decisions to be made that are almost too daunting to contemplate in a mere multi-decade time frame. There are social consequences too: What would the world do with trillions of dollars of useless oil wells, refineries, pipeline systems, ocean tankers, petroleum tank farms, petroleum home furnaces, petrochemical plants, the vast piles of now-useless petroleum employees? The list is almost endless, but it is reality.

However, just as it will be difficult, it is eventually going to be necessary. Since the beginning of the industrial revolution, we've leapt from one available energy source to another—wood, coal, oil. Each colossal wave of human development and population growth was fueled by these advances, and each brought reliance on these successive dominant energy sources.

INTRODUCTION

Every source is exhausted eventually, though. The age of oil will wind down as well. There are debates as to whether the dependence will be lessened due to climate change or to availability of reasonably-priced resources, but either way, the future holds a massive energy transition.

The topic of climate change is of course part of the story, and I'm aware that is a ridiculous understatement. Climate change *is* the story to many. It is driving massive political upheaval and rewriting government policy. Activists working to prevent climate change come from a certain angle, one that is not particularly concerned with functional difficulties. This view deserves examining. The world has gotten into plenty of trouble before by visionaries who attempted a new world order without considering consequences. We would be dumb to allow that to happen again without a lot more scrutiny.

To understand where we are with respect to any movement away from fossil fuels, it is necessary to explore the role of cheap energy in our current standard of living. No matter where you stand on the spectrum of energy debates, there is no getting around the fact that fossil fuels have enabled the most magnificent standard of living the world has ever seen for billions of people. In fact, those billions wouldn't exist without fossil fuels.

Millions work in the fossil fuel industry, and being newly-designated as climate destroyers is unnerving to say the least. A great many of these people count themselves as hard-core environmentalists and don't find that irreconcilable at all. Almost none are indifferent to the environment, much like in any other business. There are some, but there are jerks in every walk of life. For most in the energy world, there seems to be only two sides to how these people feel or react to the vitriolic anti-fossil fuel messages. First, the workers take it personally; no one likes to be accused that they don't care about the environment. They simply observe that the world needs petroleum and needs it badly. Second, businesses have their own distinct reaction: capital moves. It doesn't protest, it votes with its feet. There is an erroneous assumption that there is a fight between the petroleum industry and environmentalists. It's true that environmentalists may want to be rid of fossil fuels, and it's also true that certain petroleum industry members are fighting for their business lives. But the big money doesn't care; it just moves on to something else.

THE END OF FOSSIL FUEL INSANITY

* * *

This book is not about politics or religion; those are intractable and useless debates, for me anyway. I am interested in multiple points of view on topics that can be resolved, and I believe that the energy wars have a viable solution. They need to. How can rational people be so vehemently opposed when discussing the physical realities of what keep us alive? But at the same time, the fossil fuel debate needs to be distilled; that is, we must make sure it doesn't become a political or social engineering issue. Many people believe it is, that fossil fuels are intertwined with social justice, but if we allow the critical issues at hand to slip into that ungodly realm, there is no chance we will make any progress.

It's sad to say, but necessary: The energy world is complex beyond belief, and the lack of knowledge is a big problem. Of course, it is not necessary to understand everything, but it is imperative that people fully understand the magnitude of our fossil fuel reliance before coming up with schemes to end it. You don't get an alcoholic to quit by hiding his secret bottle of vodka, and we won't get people to quit consuming fossil fuels by preventing the flow of cheap energy into their lives (by blockading infrastructure development, for example). We need a more substantial discussion, and to have that we need to understand the fundamentals of how it all works.

A large part of the problem is that the fears that drive both sides are miles apart. Those that fear climate change and its impacts see a looming end to life as we know it, and the only solution in their eyes is to stop using fossil fuels. The other end of the spectrum observes a global population that requires incredibly large quantities of fossil fuels for survival, and sees a strong and combative campaign to wipe its business off the face of the earth. This has a personal impact, in that the industry provides many jobs. It also has a philosophical aspect: How exactly will the world survive without fossil fuels? What would it mean to the world to blow up the fossil fuel industry as the eco-fringe would like? Does anyone understand the consequences of that?

We need to build a bridge between those who fear climate change will destroy civilization, and those who think the fear of climate change will do so sooner. The bridge can be built, but only after we've made a serious effort to understand what those other people are so frustrated about.

INTRODUCTION

* * *

You might be wondering about the author at this point, and from which vantage point he speaks. I began writing about energy issues five years ago, answering various questions posed to me by friends who thought I could answer because of my background (I have over twenty-five years of varied energy experience). For whatever reason, people felt comfortable asking me what fracking was, or if pipelines are safe, because I've had experience in many of those aspects of the business. I've been a financial analyst, accountant, corporate spokesperson, communications director, executive, and crude oil trading analyst, among other things, and my answers seemed to resonate with non-energy people. One reason for the resonance is that I grew up on a farm speaking the same no-nonsense language that the occupation is famous for (although a little nonsense here and there makes it a lot more fun). I realized there were many who simply wanted to understand the business, but did not find help in a digestible way. From that start came a personal energy-related website, then a column in a commercial energy website, and then this book.

The purpose of this book is to further that objective, to help fill in the blanks for curious people. It is to defuse and deconstruct the current quagmire, where we are urged to abandon fossil fuels as soon as possible in order to save the planet at the same time that every single one of us can't even begin to estimate what it would be like to live without them. From the number of comments, questions, and interesting exchanges I've had on the subject, people genuinely want to understand. They want to learn, and not from ideologues on either side who shine light on only half the debate, and crap on the other half.

I get many pleasant notes from readers asking how to get the core messages I write about out to a wider audience. The readers of my online posts consist largely of energy people, so this book is aimed at my audience's audience, a perspective for those who want to learn more about the current fossil fuel wars and how we got to this antagonistic, panicked state where to live is to be a hypocrite.

I will shamelessly borrow a concept from Nassim Taleb, who wrote that he prefers that imperfections be left in his books as a reflection of his imperfect character. That's music to the ears of this lazy writer. But he has a point. This is a book about ideas, not about machine language or rigid thinking. It is about pushing the envelope of thinking, to push out the walls of the debate. As such, these are steps on ground that currently has few footprints, and there may be missteps. There may even be a typo. There are sentences some deem as too long, but they are that way for a reason (I like them, and they sometimes say things differently than a bunch of little sentences). Months could have been invested to resolve all these in a manner satisfactory to all editors, but my preference was to have this the book on the shelves in this crucial time of environmental debate.

This book will not be for everyone, and that is perfectly acceptable. It is meant to appeal to most, not all. One senior petroleum industry veteran told me it is insulting to the oil business; an environmentalist-type told me the book has few rational arguments. I am quite pleased with that balance. I also expect those themes to continue in their respective worlds, for the same reason socialists and capitalists don't jump back and forth into each other's camps. The book may be insulting depending on which seat you're in, and it may seem logically inconsistent to those who don't recognize the mother of all false dichotomies that is being thrust at us: that we need to get off fossil fuels right away or we are all doomed. That last bit is why the book is here, not to repudiate that position but to point out that no one really knows how hard it will be to accomplish that. I am therefore fine with having no friends in either of those camps, because I also had two other reviews: one had "LOL" written in the margins with pleasant regularity, and the other was for an order of books. Those four reviews give me great comfort that this book is not far off the mark.

* * *

We are now at a cataclysmic confluence of forces that might just make us consider these challenges. That is a reference to our energy crossroads, with one road pointing towards the status quo of a fossil fuel-based future (and total planetary destruction, according to some), and the other towards a purely and necessarily green energy future (which is fantasy and unattainable, according to others).

Neither of those views is right, and there is an alternate path—we just need to find it. It's time for a new look at the energy world.

1.
Energy as we (don't) know it

The public's understanding of the energy business is almost non-existent. Beyond the price of gasoline, almost no attention is given to energy, and the messages the mainstream media pounds out on the topic resonate less than a twelve-hour podcast of someone reading the dictionary.

This lack of knowledge is nothing to be embarrassed about. Virtually no one understands it completely. The same is true for a lot of businesses, but few are so complex, and yet lumped together in one bucket as simply "the fossil fuel industry." If we think of a typical cellular phone and the industry that makes it, we think of Apple and a bunch of cool people hanging out with stacks of money laying around like bales of hay. The phones seem to magically appear from China. We don't think of the mines, aluminum smelters, and countless other industrial processes that go into that phone. And that's fine; an aluminum smelter really isn't part of the "phone business," even if it is a necessary component of the chain.

In the energy world, it's different, because all the components really are part of the fossil fuel industry. From the people that find petroleum down the chain that gets to your fuel tank or furnace, all of that tends to be summed up as "fossil fuels" and viewed as a single entity. There are countless subindustries involved in the production, processing, storage, and transportation of fossil fuels. Each of the big three—oil, natural gas, and coal—is a massive industry in itself. Most people on the outside think of oil as oil, which arrives in one simple motion when a super-rich baron or sheik drills a well and whips up some gasoline, and then puts it into your car while simultaneously siphoning all the money out of your jeans. We hate them all because it's right there in our face, every dollar they're taking out of our pockets to stuff in theirs. We don't know how the gas got there and we don't care, but if I could get my hands on one of those scrawny sheiks with two hundred Lamborghinis I'd... But then we drive off happily to do whatever we do that would be impossible without fossil fuels, and we forget all about it. We wouldn't change our way of life for anything.

What's different about the fossil fuel industry is that people are trying to kill it. What makes that plan so unhelpful is the extent to which we rely on fossil fuels in an incomprehensibly large way. The world doesn't need to understand how a refinery works, but to have a reasonable chance of transitioning to green or renewable energy, it needs to know how much it relies on fossil fuels as a starting point.

There are thousands of examples, but one will suffice: natural gas to heat homes and buildings. Much of the world's population lives in colder climes and is subject to winters. All those billions of people suffering from this seasonal condition need heat to survive, and fossil fuels are, at present, the only game in town. Burning wood or other flammable material is second, but billions of cold people would burn every combustible item in sight without fossil fuels. Wind and solar provide almost no help with regards to home heating, and of course solar power is significantly less in winter anyway, when the sun barely climbs over the horizon.

Given those constraints, try to imagine the task of keeping all of humanity from freezing in winter, and while you're at it, try to imagine the infrastructure required to produce, process, and distribute all that natural gas. When you're growing up, if you're exposed to the demented notion of a Santa Claus that actually delivers presents to every child on earth, you quickly grasp that a single out-of-shape elderly person isn't likely the distributor of all those presents. But with natural gas, a reasonable facsimile exists: an inconceivably large network of pipes, from large to miniscule, that delivers right to your house the exact quantity of natural gas that you need to keep the thermostat within one degree of where you like it. Remember also that natural gas is not like a dump truck full of coal; the entire system must be sealed to perfection, and any leaks out of the countless thousands of miles of pipe will result in either a nasty smell or an explosion, or an even more bombastic lecture from Greenpeace. The scope of the natural gas system alone is a breathtaking achievement, which ironically will never receive even one-billionth the attention of a Hollywood star that has gained three pounds.

On top of these examples is the reality that much of the world's population wants to enjoy the same standard of living the west does. Actually, they'd probably settle for a quarter of it, but to get even that far will require stupefying amounts of energy that, for at least the next few decades, will massively rely on fossil fuels. In 2017, two years after the Paris Climate Accord was signed to many global enviro-celebrations, 1,600 coal-fired power plants were under construction.[1] Countries like Canada and Australia valiantly shut down coal fired plants,[2] single digit totals, despite (in Australia's case) causing grid instability and rolling power blackouts, and at astronomical cost. Simultaneously, sixty-two countries were constructing roughly a hundred times as many coal-fired plants as were shut down in these two examples. Symbolism can be expensive and futile. In addition, the narrative that we think we know about energy is often dead wrong, because there will be a thousand headlines about new solar installations (and the potential energy they can produce, not the practical amounts) and zero about the world's new coal installations.

As the developing world sharpens its consumer skills, global oil consumption continues to climb relentlessly; media reports often erroneously imply demand is falling if the rate of growth slows. Automotive density in China and India pales in comparison to the western world, and their citizens are not gloriously happy to maintain that disparity. They are buying vehicles hand over fist. Some are electric and many are small, but all add up to new energy consumption.

These trends tend to cause panic in certain circles here in the west. Some have adopted the attitude that "we got ours" and the rest of the world can do without because the environment needs saving immediately. But upwards of 4 billion people don't "got ours" and they really want it. We can't blame them.

These are all events of significance in the world of energy, but we don't see them. We hear that green energy is taking over, that fossil fuels' days are numbered, that activists have successfully campaigned huge investment funds to avoid fossil fuel investments and dump shares. We hear nothing but that, and we are getting the wrong impression in a major way.

Some energy market commentators try to explain energy realities with buckets of statistics and charts. That's mistake number one, as far as the public is concerned. We're not all engineers and accountants. When was the last time you met a friend for coffee and showed each other charts or data tables? In business or academics, a chart can be worth a thousand words, but there are a lot of people out there that don't work that way. Fossil fuel advocates fight the fear-based messages of impending climatic doom with statistics about how much of it we use, which turns out to be quite an unfair fight. A picture of an oil spill on social media trumps a hundred PowerPoint presentations. The fossil fuels battleground is for the hearts and minds of the silent majority, and if all it took to win them over was show them some charts, this would be a wonderfully simple world.

People have long tried to engage the population about the depth of our reliance on fossil fuels. Various analyses and books offer magnificent statistics that lay it all out. For a slice of the population, that was extremely valuable, but for the majority, it is like drinking from a firehose. At the end of the day, these sources still don't explain the most pressing concern of the average consumer, whose line of inquiry ends at the fuel pumps as he/she fumes while filling the gas tank and wants to know how those damn oil companies are screwing them. This book won't help with that raging, but it will hopefully be more relatable than descriptions of how many swimming pools of oil our current consumption would fill, or other similar attempts to convey a sense of scale. Even quoting statistics in barrels is pointless; not many people measure in barrels, not even world-class alcoholics who transact and consume in precisely those units. By the time readers delve far enough in to find (and visually imagine) statistics on swimming pools full of oil, the discussion of global consumption levels has dissolved into a battle between boredom and repugnance.

One way to put in context what fossil fuels have done for us is to consider all the people you know. Of all those people, probably 90 percent would not exist or have ever existed without fossil fuels. I have no idea if that's the right number, but it is far closer to 90 percent than it is to zero, and without fossil fuels—cheap and plentiful supplies to burn for heat and power—we would have gotten nowhere industrially, and none of this would have happened.

ENERGY AS WE (DON'T) KNOW IT

No cheap manufacturing processes, no mass manufactured goods, no big buildings or hospitals, no pavement, none of it.

Before fossil fuels, the only means of creating almost anything using raw materials—for example, iron—was to burn something to generate heat for the process. That worked well enough with few people on earth, but population growth would have quickly denuded the landscape of any trees. I don't know what population the world could bear if we all lived on manual labour and by burning trees to stay warm, but it sure as hell isn't 7 billion people.

Here's a key part that few realize when eyeing a green future: there would be no solar or wind energy without cheap fossil fuels. Windmills have existed for a long time, but when simply made of wood with no bearings or electric components, they would have been relegated to lifting a bit of water, grinding wheat, or similar tasks that a donkey could do. They would have done little other than attract lightning. Solar energy's contribution would have been to heat black rocks so we could sleep on them.

Having said all that, it is important to impart a feeling for what we require and consume to maintain our standard of living, and to contemplate what that means if the rest of the world catches up. Only with an understanding of that realm will we be able to contemplate a conversion to different energy sources on a global scale.

Let's park in-depth discussions about quantities; they don't really matter for reasons we'll get into a bit later. Who can picture a hundred million barrels, which is our daily oil consumption? There is no statistic on how many coffee cups or swimming pools or blenders that could be filled with today's consumption that would mean anything to the average person, because we don't think in those terms about anything. We don't wonder how many barrels of ketchup the world consumes, or beer, or even water. At best, those facts are interesting trivia, and at worst, non-interesting trivia.

We can begin to imagine the scope of our consumption habit when we consider, for example, a cell phone. According to a technical website, an iPhone is comprised of 75 of the 118 elements on the periodic table. The human body has thirty. I know, I know, that's a statistic and a bunch of numbers, but it's pretty wild, hey? Now, imagine the challenge of securing supplies of all those weird substances, from all over the world, to get them to the right place in the right quantities at the right time, and make a cellphone.

Now, think of the similar effect of creating TVs, food, computers, bicycles, vehicles, and on and on, and we begin to get a feel for the mind-numbing array of logistical chains in operation. The whole thing works for one reason: because we have cheap energy.

We can now flip around to the other side of the equation and imagine what it takes to get all that oil, natural gas, gasoline, and diesel into the right place at the right time to enable the manufacturing process. We might as well come right out and say it: petroleum is not the most pleasant stuff as it comes out of the ground. Actually, it's not all that pleasant when refined either. A diesel spill on your pants will remain as such for a long time. You wouldn't enter a restaurant with one. Those aspects are, of course, low on the list of why fossil fuels need to be handled with great care. They are obviously highly flammable and don't mix well with the environment (although it must be said that both are natural substances, and do exist in perfectly natural situations—methane (natural gas) from swamps, oil from near-surface deposits that can and does enter rivers naturally, etc.).

Regardless, fossil fuels need to be transported in enclosed containers like trucks and pipelines, and because of the potential for dangerous incidents, they must be handled with great care. It is one thing when talking about something small and relatively benign like a can of paint, but quite another when considering the quantities of fossil fuels that are the lifeblood of pretty much the entire world. Thus, the scope of the equipment and carriage vessels in the global fuel delivery system is truly mind-boggling (as is the chance for an accident).

Working back a step further, we don't think of where all the oil comes from either. There are the cliché answers like Texas, Saudi Arabia, and Alberta, but in reality, a better question is, where doesn't it come from? That's not to say that every spot on the globe contains petroleum, but the search area has expanded pretty much as far as it can. There is almost no point on the globe that hasn't been poked to see if oil will come out, the only exceptions being locations where it is physically impossible, like mountain tops, the Mariana Trench, and any place in close proximity to the residences of either environmental activists or corporate executives (at least they have something in common). Wells are drilled into the ocean floor that are so deep and complicated that the cost of a single one can exceed $200 million,[3] and current exploratory wells are now drilled in the middle of nowhere. For example, Brazil made a significant discovery a few decades ago that was 170 miles offshore and more than a mile below the surface. One of the latest major oil developments to come on stream is located in the Caspian Sea offshore from Kazakhstan, costing in excess of $50 billion.[4] You can be sure that if companies are willing not only to go to Kazakhstan, but offshore Kazakhstan, that easy-to-find oil is no longer part of the equation.

None of these things come to mind if we only think of fossil fuels in terms of filling up the car or heating the house. To make matters worse, when we are curious and pay attention to news feeds, we seldom find useful information. Even the commentators designated as experts, the ones that dominate the airwaves with oil market commentary, often don't understand the business. They act like they do, and they sound convincing when they bark out ludicrous comments like "oil up two percent today on geopolitical tensions." It's all gibberish; prices were up or down because traders were playing games and made bets based on how they guessed that the talking heads would interpret the move. Then the whole cycle of dis-informational feedback convinces you that you're a moron and they're all experts, and there's no way to understand what's going on so why even try.

It is worthwhile to try to understand the energy business though, for reasons beyond soothing the irritation that arises when filling up the SUV. There is an energy war going on, made up of opposing camps that appear to despise each other. One of these camps is quite content with the reliance on fossil fuels and works to maintain the relatively comfortable status quo. The other side is not happy with your dependence, and wants not just to send you to a counsellor, but to slit the throat of your supplier.

There are fundamental flaws in each side's positions. The fossil fuel advocates are running out of cheap reserves, meaning the world will need to find an alternative sooner rather than later anyway. Talk of a hundred years of reserves is, based on consumption patterns and global new petroleum discoveries, gibberish as well.

THE END OF FOSSIL FUEL INSANITY

For those that fight tooth and nail against fossil fuels, their position is also problematic because they are consumers just like everyone else. They have no ability to sensibly articulate how to live without fossil fuels. Some will float hallucinatory and comically unrealistic plans to abolish, say, internal combustion engines within twenty years, plans which have just enough of a whiff of possibility about them to catch your attention. But the notion of changing entrenched systems in any reasonable timeframe is, under any microscope at all, preposterous.

2.
How did we get in this mess?

PART I: Systems developed around fossil fuels

LOST IN THE GOOD LIFE, WE HAVE NO RECOLLECTION OF HOW WE GOT HERE

My goodness, look how you've grown. You're what, two hundred now?

That's a reference to the industrial revolution, and unfortunately, the energy saga requires a fair bit of historical context. History, for me anyway, is annoying, because it tends to be, with the exception of hard physical realities like "Stalin's pet died in 1932," largely just biased interpretation. If one were to read about the demise of Stalin's dog in a historical text, the retelling would surely include some subjective interpretation such as "Stalin's beloved pet, most likely a Chihuahua according the memoirs of his former housekeeper, died in 1932 from what are believed to be multiple gunshot wounds that researchers attribute to the dictator's growing paranoia." Fortunately, we don't need to get into interpretation too much. We can hit the high notes to trace what fossil fuels have done to provide us this cushy life.

However, it is necessary to give a bit of historical context as to how we got in this mess, because without it we fall into the overwhelmingly common psychological trap of thinking the world consists entirely of what we see of it. This is not revolutionary thinking; the fable about the blind men and the elephant captured it all eloquently long before Stalin's dog was unfairly judged and executed. No matter how much we travel, it happens; our tendency to assume all worlds are similar to our own is strong. We think that what we aspire to is what everyone does. We assume that dogs are beloved pets everywhere in the world because we love ours.

We also assume that everyone's world works as ours does, that if you want water you turn on the tap, and if you want electrical power, you do nothing at all because it's right there in the nearest wall. At present, for example, sitting in the middle of North America in a comfortable chair, enjoying a standard of living unlike the world has ever seen before, it would be natural to conclude certain things about how the world works: that we've got it all figured out, that we are masters of this universe, that we understand the planet and the climate, that we need to change (and are changing) immediately to green energy, and that it's happening as we speak without so much as breaking stride. The inexorable message in the current hyperkinetic minute-by-minute news feeds is that the industrial revolution is somehow over. Well, not quite over, but stabilized. Surely, we've entered or are entering a panacea of endless free renewable energy that will ensure forever that my lattes will be delivered to my insatiable self exactly as I like them, whenever I want them, or the world will face the wrath of humankind's current most important person: the indignant consumer. True, some fear the planet will warm too much and threaten our world, but even those predictions (except for the outlandish ones) discuss a changed world of higher sea levels, flooded coastal cities, and different weather patterns. At the core of even these scenarios is the fundamental belief that everything else will still be there, all the millions of wonders that make up our daily lives.

Before discussing that cultural phenomenon, we need to put some context around the systems that have been developed in conjunction with the industrial revolution. We need to examine the things that have transpired to make the miracles of yesterday commonplace today.

VERY BIG SYSTEMS DEVELOP WEIRDLY AND (NEARLY) PERMANENTLY

Have you ever been to a very new city, or to a very old one? Both offer excellent case studies on the development of systems. The most interesting ones, though, are cities that are forced to adapt to rapid change (rapid as in a few decades, not rapid as in a tornado).

I happen to live in a new city, historically speaking. Calgary, Alberta, has grown from a small outpost to a city of well over a million in just a hundred years. It has therefore developed entirely within the envelope of the industrial revolution, mostly within the past seventy years or so, after the automobile had relegated to curiosity status all other forms of transportation such as things with hooves or two wheels (although middle aged men on Harley Davidsons are making a pseudo-rebellious comeback). It therefore offers an interesting petri dish in which to observe how major infrastructure systems develop with regard to existing and planned constraints in a not too old, not too new city—a fitting parallel for what we face in rewiring energy systems.

Consider first how ancient cities adapt to new technology. It seems that cities that develop organically over centuries are pretty much set that way, that there is not much that can be done to adapt five-hundred-year-old horse trails that became streets into modern architecture and infrastructure. The solution then is to simply build over or around these ancient cities, because to adapt them would be to destroy them. Paris is an excellent example; the city has developed a ring road and a series of overpasses that simply bypass the centuries-old beautiful mess below. There is no effort made to make the center of the old city capable of handling huge traffic loads. Paris was massively redesigned[1] in the 1800s; broad thoroughfares and parks were added through the "demolition of medieval neighborhoods." That project was controversial at the time and work on it continued from 1853 to 1927. If it took seventy years then, it would take two hundred now.

Calgary is a "tweener" city, old and developed enough that new traffic infrastructure can't be designed from scratch, yet young enough, and growing enough, that solutions must be found to make old systems work in new ways (albeit with much difficulty). In ancient cities, there is no longer any question of demolishing old buildings or blasting open neighbourhoods for new thoroughfares. Those old habits don't cut it any longer.

A city like Calgary illustrates the extreme difficulty of rewiring established systems. Fifty years ago, the city built out transportation solutions that were anticipated to meet the needs of a growing city. They did for forty years, but now, growth has continued, and that established infrastructure is a mighty thorn in the side.

One can see this readily if they are brave enough to drive around the city. It is an abomination to get around by car because the urban planners of half a century ago laid down big building blocks—the main traffic arteries—that no longer work, and there's nothing to do about it now. These days, an emperor like Napoleon cannot simply wave his hand and say, "make it so"; the consultation process alone would have forced his capitulation. Indeed, Calgary has no emperor at all. What it does have is patchwork fixes that dance around the modern needs of consultation and expense. The result is chaotic; thoroughfares traverse a third of the city, then come to dead ends just when you think you can use them to get somewhere. City planners block off certain streets to all but public transit through ingenious "vehicle traps" that torment newcomers and tourists who can see where they want to go but find the route blocked (because it was too popular). Certain neighbourhoods can't be entered into at rush hour because frustrated citizens at one time tried to short-cut through them, and the police will not stand for that. The city has basically one major east-west traffic artery (one other, the mighty Trans-Canada Highway, is festooned with more traffic lights than Sunset Boulevard has palm trees). The city is littered with bizarre interchanges, brainlessly disjointed thoroughfares, mystifying traffic signs, forests of perfectly mis-timed traffic lights, and a plethora of moronic subdivisions that make many neighbourhoods virtually unvisitable and yet paradoxically inescapable.

Pedestrians fare no better and often much worse. Traffic routes are designed by young guns at CAD machines who create bizarre new neighbourhoods and street patterns straight out of a textbook, some academic's vision of what a modern subdivision should be like—i.e. car-centred, ultra-safe (for cars), utterly sterile, and builder-friendly. Thus, the modern pedestrian is faced with jockeying across landscapes developed entirely around the needs of the driver, with commercial infrastructure sequencing dictated not by the needs of the citizen but that of the planner, and baked in forever. A coffee shop or sign of humanity has no more chance of popping up in a suburb than does a pig popping up in Buckingham Palace. Going for a walk in a modernly designed urban development is nothing but an exercise in frustration and death-defying pedestrian crossings, with sidewalks beginning and ending randomly, and no way to get to where you want to go except by following the migratory but militarily enforced auto-friendly traffic routes and hoping you don't get run down on the way. Not that I'm bitter or anything.

This litany of complaints illustrates not just my sadly short fuse when forced to deal with criminally horrible urban planning; it also shows the challenges of building new infrastructure within existing structures on a large scale. A blank sheet of paper would be a joy to work with from an industrial perspective because there are few impediments to optimization, but those are truly rare opportunities.

Of course, many of the world's mega cities have not developed to the same wildly exorbitant lifestyles we enjoy in the west. Many of the world's city dwellers—and the global trend is relentlessly towards urbanization—live in huge cities where the poorer segments have developed crude systems that work, more or less. Countless power lines run from shaky posts to balconies, to rooftops, and beyond. The systems may be less polished, for example with fewer traffic lights, more insanely hyperactive intersections, or highly visible but tourist-discouraging sewage systems, but they function quite well considering the number of people crammed in.

You might be wondering what all the amateur social geography analysis has to do with energy. The answer is nothing directly, but indirectly, everything. Just as Calgary has extreme difficulty in adapting to a growing population from a transportation perspective, the world is now being asked to rewire itself to handle an energy revolution. The comparison to remapping a city is not ridiculous when one considers the scope of the fossil fuel-based infrastructure that keeps our world turning. The task would be monumental.

MODIFYING OR REPLACING SYSTEMS IS UNBELIEVABLY DIFFICULT

Remember Ross Perot? Small guy, a Texas billionaire that at times resembled a plucked chicken? He was actually a remarkable businessman and visionary that made a decent run for the US presidency as an independent. He also provided us with one of the more memorable and graphic examples of the difficulty of changing large, entrenched systems, even when leaders wanted to.

Perot founded a company called Electronic Data Systems, which he sold to General Motors for $2.4 billion in 1984. When GM took over EDS, they liked Perot's feistiness and no-nonsense attitude. They thought he might be good for GM, that he might clear out some cobwebs and bring a new sharpness to the behemoth's culture. The considerable size of the buyout and Perot's obvious business acumen made him particularly attractive to GM's executives, who placed him on the board of directors in hopes that he would infuse the company with a dynamic new spirit.

That he did for two long years. He hit the ground running, and almost instantly began "helping" GM understand where their bloated and obscenely large structure needed updating and streamlining. His suggestions were common sense, wise, and achievable, but they were also difficult. To the GM board, it was initially refreshing to hear Perot say such things; it was another entirely when he tried to have them implemented. His straight talk and bluntness began to wear thin as he talked about the need to "nuke the GM system." In short order, Perot went from being an interesting commentator at the board room table to an angry bird circling GM's management, dive bombing them like they were a nervous dog that got too close to a nest. They talked about how Perot needed to understand GM's system, and he told them to shut down the executive dining room and park with the rest of the staff.[2] GM management, who had changed their mind about wanting an overhaul, booted him out the door with a huge payout, and even that didn't shut him up: "It is inconceivable they should spend this much to buy me out," he commented in an interview. "[It] is enough to build a state-of-the-art auto plant." Even in mid-air with a distinct corporate boot print on his bum, he continued to spout the kind of common sense wisdom that GM's system simply could not handle. GM famously went bankrupt a few decades later, unable to adapt to the realities of the modern auto business.

GM also provides another excellent example of the difficulty of modifying systems. After the fuel crisis of 1973, when the US ran out of gasoline due to an oil embargo from Arab nations, US auto manufacturers were required to substantially increase the mileage their cars achieved. GM had the chance to license technology developed by Honda, a new type of combustion process they were using in their little Civics, but GM's CEO mocked the technology, commenting "While it might work on some little toy motorcycle engine… I see no potential for it on one of our GM car engines." Honda was miffed at the insult, and the much smaller Japanese manufacturer took the opportunity to give the auto giant a steel-toed boot to the crotch. Honda purchased a GM-built Impala, shipped it to Japan, and custom-built engine components to prove the technology would work on GM's own vehicles and could meet new mileage standards that GM said was impossible. GM had other ideas, including one preposterously stupid one: they chose to convert one of their gasoline engines to run on diesel, and set some sort of record for one of the worst ideas in manufacturing history as the modified engine failed at a prodigious rate.

Of course, this example is made more emphatic by the old GM's arrogance and complacency; nevertheless, it is an excellent example of just how difficult it is to modify large, existing systems. There are few shortcuts.

Here's another example of an attempt at a large-scale system upgrade. The government of Canada is in a pickle as this is written. This problem is with the government's ability to pay its own people. Part of the government has been paralyzed by a new computer system, and the mess is so bad the government has in some instances been unable to pay its staff for some quite-ridiculous periods of time.

The government, several years ago, implemented a new payroll system dubbed Phoenix. The system does not follow in the footsteps of its mythological antecedent namesake, which exists in a cycle of dying and rising again from the ashes; the payroll Phoenix is like a drunken one that rises every day and flies straight into a ceiling fan and crashes to the floor, in an infinite loop.

This is an interesting example for several reasons. First, it is almost all software related, and yet look at the chaos that alone has created: years and counting of dysfunction. The amazing part is that this was a change to one subsystem of a complex government operation. One of the government's other main systems, the revenue collection system, is of a 1960s vintage and so completely band-aided and ancient that the government is too terrified to even contemplate replacing it. As pathetic as it sounds, it's one thing to go without paying employees—those are nothing but a bunch of tiny tragedies—it's quite another if the government can't collect its taxes. Now that's a problem.

If the plan is to rewire the world and make it a smart grid where power can be generated anywhere and fed into the system, the software upgrades alone that will be required to electrical systems will be profoundly difficult. As we will see in a subsequent chapter, upgrading antiquated power grids to today's world is like getting a horse to use a cell phone.

As a simplistic overview, the power grid been designed as a one-way system. It connects multiple large-scale power sources together, moves that power at high voltage, and then breaks it down into ever tinier subsets until it reaches your Tesla or hair dryer in relatively non-dangerous amounts.

It is actually a fragile thing, far more than we would expect given our reliance. If a single storm or terrorist attack brought down a major transmission line, there would be havoc on the grid for a great distance.

With the rise of renewable energy, the plan is to use the micro-sites (individual dwellings, rooftops, etc.) not just as users of power, but providers as well. The system is not built for that. It is going to take a remarkable amount of software and hardware to reconfigure the power grid to allow large-scale changes to how it operates. Not only will we have to build new systems, we'll have to dismantle old ones. This point is almost never addressed, but when considered in its totality, it is a massive problem. Changing widespread and established systems is incredibly difficult.

With regards to fossil fuels, what will we do with all the outdated equipment like tanker trucks, pipelines, and distribution terminals that would be left behind if fossil fuels were abandoned quickly? That is only one aspect of this issue. What about the millions of people put out of work in high paying jobs? What about the hole in the pension plans of almost everyone on earth that has one? The market capitalization of the top oil companies is in the trillions, and they pay a lot of dividends. To abolish those entities may cause dancing in the streets for a small slice of the world's population, but will cause great consternation for almost everyone else.

Already I'm getting into dangerous territory because some will see the preceding paragraph as nothing but a defense of fossil fuels and the status quo. It could be construed that way, but it also reflects realities that anyone in any position of authority will have to consider. In some ways, it is simplistic to fight for the demolition of something and offer theoretical possible solutions in its place; it's another entirely to be responsible for those solutions working correctly.

LAST BUT NOT LEAST: THE HUMAN SYSTEMS

There are two fundamental types of systems: physical ones and human ones. The first is profoundly hard to change; the other is nearly impossible.

Physical systems have one characteristic that makes them slightly easier to change: it's mostly a question of money, and to a lesser extent, time. Money can work wonders. Don't like where a power plant is? It can be moved; all it takes is money. I didn't say it was easy or cheap, but it can be done. With enough money, it can be done quickly. Not like three days, but expeditiously.

Human systems are far harder. Professions build up over time, with structures around them, and humans become experts in these fields. Some, like medicine, simply build on past knowledge and it all works wonderfully. Others are more abstract, even if they don't always appear that way. Consider the accounting profession. It is built on a set of reporting standards that have developed over decades. In North America, there are close to a million certified professional accountants. Each has spent significant time becoming experts in current reporting standards. The financial world works on these standards, yet accounting theorists (yes, there are such people) view the current system as outdated and overly-patched, and argue that it yields ever-more useless information—for example, using the "historic cost" principle to value assets acquired fifty years ago. The old framework is not designed to work with the modern world, yet it will never change because of those millions of accountants that keep it going.

To throw it away and start again would mean, among many other things, invalidating what this entire profession is good at, and asking them to help develop a replacement.

Who does that? No one does. No proud professional with decades of experience stands up and says, "What I've been doing for twenty years is a total waste of time and I vote we toss the whole thing out and start again." No petroleum professional stands up and says, "Yes, I'm responsible for destroying the environment and causing climate change and putting my

THE END OF FOSSIL FUEL INSANITY

kids' future at stake." No environmental activist stands up and says, "I'm as much a part of the problem as anyone because I fly wherever I want whenever I want and I won't go without my fair-trade coffee every morning."

Any sizeable system is hard to modify, and the fossil fuel one is as big as it gets. It impacts how billions of people eat and survive, and even where many live. In the news, we sometimes hear, for example, projections from "futurists" such as economist Tony Seba who claim gasoline cars will no longer be sold within eight years[3] (and he made the claim a year ago). While the sensationalism of the message gets these stories into the headlines, it is extremely hard to reconcile such projections with reality.

3.
How did we get in this mess?

PART II: Overreliance on fossil fuels

When talking about an overreliance on fossil fuels, the best place to start, as one might expect, is with bananas and pizza. Between those two culinary staples, we can paint a clear picture of how we got to our state of dependence.

Let's start with the intercontinental banana syndicate. In addition to being a vital component of juvenile humour, bananas provide a fascinating example of how reliant we are on our present industrial systems, including fossil fuels.

Consider two aspects of this noble fruit, whose reputation is sullied not just by cheap jokes but by an overly emphasized monkey association. Bananas mean more to us than they do to your average monkey. The fruit is an incredible staple: grocery stores go through them by the ton, and despite that relentless demand, it would be a wild day indeed that saw a modern supermarket emptied of them. Americans eat more of them than any other fruit, some twenty-five pounds per year[1] or approximately a hundred bananas apiece. Canada is even worse, if consuming more equates to worse, at about seventeen kilograms, or thirty-seven pounds, per hoser per year. Apples are second and not even close.

On top of our propensity to drown our sorrows in bananas from an early age, consider the second noteworthy aspect of this fruit: the remarkable journey it makes to get to your shopping cart, and how excruciatingly fragile they are as cargo. Nothing is as delicate as a banana, not even a baby. If you're pushing your infant around the grocery store in a stroller and set the jar of spaghetti sauce on junior for a second until you can get to the till, as long as your spouse isn't around, all will be well, and there will be no lasting damage to your bambino. Try the same with a bunch of bananas; if you place them at the bottom of the bag and the spaghetti sauce on top, you'll hear about it for fifteen years, and the bananas will be destroyed. Now, consider that bananas can only come from tropical countries and have to be handled with remarkable care from the moment they leave the tree. They are transported at temperatures between 13.5 and 15 degrees Celsius (56.3 and 59.0 degrees F), a temperature band that is almost preposterously narrow, and are ripened in special rooms in the destination country that are filled with ethylene gas to promote proper ripening.[2] A banana is more coddled during its life than a cruise ship passenger.

Even then, once they've completed their expensive and eccentric journey to your table, they have a shelf life of about three days before they start turning black and the kids declare them inedible and disgusting, and they head for the final resting place of either the compost bin or banana bread. Despite all that delicate handling, any urban centre in North America, including the cold ones in Canada or Alaska, can provide you with a handful of bananas any day of the week for far less than an hour's wages.

Bananas are an example of the phenomenal power of our delivery systems, but also the depth to which we've become dependent on them. Imagine a city without bananas; it would be bedlam. Okay, maybe that's an exaggeration, but we'd be pretty freaking annoyed. We simply aren't used to living without these sorts of things.

Bananas are just one example, but stand in the middle of the supermarket and marvel at the selection, where it came from, and how fresh it is. Fresh mangoes, pineapples, eggs, and fish line the shelves, all appearing in perfect condition from who knows where via some disjointed and Rube-Goldberg-like delivery apparatus spanning the globe. Each cog has to work more or less flawlessly every day, or we don't get our sushi and there's hell to pay.

Without fossil fuels, that supermarket would most likely be bare, or maybe it wouldn't be there in the first place. Consider any northern city. There is no way that these places would exist as we know them, because there is no way that the land could sustain that many people through harsh winters.

HOW DID WE GET IN THIS MESS?

We have reached a state of industrial progress and optional blissful lethargy undreamed of a century ago. We can see evidence everywhere if we care to think about it, and if you don't, think about this: pizza delivery to your doorstep is an unparalleled feat in the history of humankind. Remember that the next time the delivery person shows up; they deserve more than hoots of derision at their rusty car and more than a tiny tip for rushing hot food over for you to devour while arguing over the respective merits of various goal-scorers.

Never before have we reached a state of convenience where for a few dollars someone somewhere will craft you a meal that is exactly to your liking, then throw it in a car and hustle it to your door while still hot. Almost no one thinks that is a big deal anymore, and children think of it as a universal human right. Western children do anyway, ones that have been brought up in lands of unimaginable conveniences that are as taken for granted as oxygen.

It's no wonder we became so reliant on fossil fuels. The choice is simple: give in and float in this luxurious life filled with wondrous things at our fingertips and travel where we want when we want, or join the Amish. The purity and simplicity of their lives is in some ways admirable and wholesome, but in the same way that life in a Tibetan monastery is. We all want to achieve enlightenment, but on the other hand, comfy couches and freshly delivered pizza have their merits.

One of the biggest contributors to our fossil fuel dependence has to do with the cost. Petroleum is an absolute steal, believe it or not. That may not seem apparent during a random gas-price spike where it costs a hundred bucks to fill up the family beast, but consider the path that gasoline goes through to get to your car compared to, say, a bottle of Coke. A drop of oil already has a fortune behind it before it ever sees the light of day, and it gets a lot more expensive from there. It costs money to ship it, refine it, then transport it back to wherever it is needed (everywhere) in a highly-purified state. The cost of each step is staggering. Think of the capital investment in a refinery or pipeline system or ocean tanker or drilling rig. Add on the cold, clammy hand of the government, which slaps taxes on fuel from a variety of angles, and it's a wonder that gasoline isn't a plaything of the wealthy. Consider jet fuel: for a few hundred dollars, we can be transported across the country or an ocean in the relative luxury of a modern airplane (and it is luxurious, no matter how much we complain about waiting or screening or the selection of movies). That option exists for many on the lower rungs of the economic ladder as well; at no time in history has the bottom income quartile had that sort of option. I once met a cleaning lady who flew back to Hungary every four years to see family. Earlier generations of that profession were lucky to venture across the city.

All this turns on cheap energy. It underscores everything we do. A growing argument is being made that renewable energy is filling the void—and it is making a start—but at the same time, demand for non-renewable energy is growing as well. We actually have a far bigger problem: the whole world is starting to expect daily fresh bananas too.

4.
How did we get in this mess?

PART III: Divergence of opinions

In the current fossil fuel wars, most people fall into three camps: a small percentage hate fossil fuels and blame them for ruining the climate; another small percentage either works in the industry or truly understands our reliance on fossil fuels; and the majority in the middle change the channel when the topic comes up. In the quest to get that middle camp to understand the importance of fossil fuels, it is relevant to look back at history to analyze briefly how the simple hydrocarbons that powered the industrial revolution became so polarizing to some while engendering indifference in most.

We can start the analysis by going back a bit to trace the origins of the two main competing factions. The stage was set for the divergence little more than a half a century ago, and that stage is best viewed in the context of some much older history that we can fortunately bounce across rather quickly.

Let's look briefly at the history of war. What we are concerned with here is not who attacked whom in the year 1358, but the fact that in all likelihood, it happened. And it happened in 1458, and 1558, and so on. History is riddled with insecure men (usually men) attacking each other for God knows what reasons. In fact, God is often one of the main reasons. Some sort of god anyway, and the ones who didn't bother to believe in a god were run over and mauled by either those that did, or those that were insecure, or both.

THE END OF FOSSIL FUEL INSANITY

This sad commentary on humanity continued at an unhealthy pace right up until 1945, more or less. Until then, wars were like harvest season, showing up with sad regularity.

For some reason, possibly through industrialization or because everyone was smoking and relaxed, or because the advanced weapons took the sport out of it, an intolerance for these global conflicts developed, and the world found ways to avoid them. Perhaps the current period of relative peace has to do with the fact that a world war would be a doomsday scenario, given the nuclear arsenal that is floating around the globe. Wars are kind of expensive too, and modern tyrants tend to run up the credit card in other more hedonistic pursuits. Ugly conflicts still popped up, such as in Cambodia where a band of zealots wiped out half the population in a shocking experiment in new governance techniques, but by and large it's been peaceful. The former Soviet Union and the US went right to the brink, and recently Trump and North Korea's top insecure man glared at each other like drunken barroom goons, but nothing came of it—thankfully. In fact, they actually got their testosterone-addled brains in the same room without a fist fight breaking out, but considering one of the combatants is past seventy and the other looks like he'd get tired by eating cereal, their better judgement prevailed and they had a nice conversation about whatever insecure megalomaniacs like to talk about.

Regardless, because of all this peace, the world changed and reached a critical juncture.

Through the modern framework, where global or large-scale conflicts have abated, did the world change for the better, or the worse? That question obviously sounds brutally insensitive from a human perspective, but what about from the perspective of the planet's resources? This lack of large-scale warfare is quite pleasant, but has also been a key reason that the planet's population has grown by leaps and bounds. Is it good or bad for the long-term survival of humanity that people are no longer being killed in massive waves of military adventures? The question is key to understanding the current climate change wars.

In one camp are those that argue that the world changed for the better. Peace allowed people to live longer and be murdered less, which is always nice. These non-murdered people had families, and the whole lot of them ate stuff and wore clothes and started travelling. Elvis happened, and people bought cars and drove around aimlessly at will. The baby boom happened, and the world's population jumped from about 2.5 billion in 1945 to about 5.3 billion by 1990. Life expectancies around the world soared, and today eighty is the new sixty (but ninety is still pretty old).

The energy world exploded in conjunction with that rise in population. Oil consumption rose from about 10-15 million barrels per day (b/d) in 1945 to about 60 million in 1990 (and it's about 100 million b/d now). In that period, the oil industry really hit its stride as a global phenomenon, and much of the world's infrastructure—refineries, pipelines, oil handling terminals—were constructed over this period. What's important to keep in mind is the scope of what happened; the additional demands that those extra billions of people put on the world's resources as the globe shrunk, particularly in the context of our energy-hungry lifestyle.

Think about that, about all the material goods that were demanded and consumed by those additional billions of people. Think how many diapers that is, or bacon and eggs every morning, or airplane tickets. This is all wonderful stuff, a peace dividend of sorts, where supermarkets became the norm instead of hunting for grubs and catching rabbits.

Paradise, right? Well, hold on a minute. Somewhere about the time that global population hit 5 billion, a segment of the population started wondering what impact all this production and consumption was having on the planet. With all that peace, the additional citizenry that was now flourishing without constant fear of military mayhem put a consumptive load on the world like it had never seen before. In the past, nature had dealt with overpopulation or scarcity of resources by impolitely adjusting the population downwards until a new equilibrium was reached. These days, we globally rush to the rescue of endangered citizens wherever we can. This is definitely a better situation from a humanitarian standpoint, but what about from a global sustainability one?

Through this global growth, a segment of the population began to pay a lot more attention to our environmental or consumptive footprint. Obviously, a certain segment had always cared, but until then they were simply strange people who rode bicycles *by choice*, who actually enjoyed living off the land, and who read publications like Mother Earth News. They loved articles on how to make your own clothes and compost your garden from your own toilet, and recipes to hide the taste of those vegetables from your self-composted garden. These original environmentalists pursued such activities long before they became cool.

At a certain point, though, once a pool of strange people gets large enough, it turns into a movement, and they're not strange anymore because they have a lot of friends. The environmental movement may have started well before then on a smaller scale, but what's key is that it really took off. Obviously.

One reason that the environmental movement moved beyond the primary "fighting pollution" aspect and into the realm of climate change was the development of technological devices and techniques that allowed scientists to study climate history in greater detail than ever before. This new technology led to a burgeoning profession that sought to better understand changes in climate. It made sense; until the twentieth century, temperature measurements were rudimentary in most places, if they were done at all. Anecdotes that farmers like to throw around like, "it was so cold that winter that squirrels leaping from tree to tree would freeze in mid-air," may have gotten the point across, but were notably unscientific (scientists refuse to freeze squirrels). As technology advanced into the twenty-first century, weather stations popped up everywhere, including remote places like the Antarctic, as well as random fields and farms around the world. The weather dataset exploded in size and reliability, with high-quality measuring devices taking measurements at thousands of places on the globe and replacing the outdated squirrel-standard. The rise of computer power allowed it all to be harnessed usefully.

In addition, exploratory techniques and the accompanying analysis afforded by technology allowed us to peer closer into the history of climatic events. We could therefore pull a thousand-year-old plug of ice out of a glacier and subject it to a barrage of tests and analysis that shed new light on its history, whereas a hundred years ago we'd probably have simply classified the ice plug as dirty or not dirty and used it as a weapon against the inevitable approaching bear.

With this new knowledge came observations that the climate has warmed significantly since the dawn of the industrial revolution. It's hard to say if this news itself would have ever caused anyone to raise an eyebrow, but the new breed of climatologists concluded that the rise of carbon dioxide emissions that accompanied mass industrialization was responsible for the rise in temperatures.

This finding caused alarm among climatologists—not all of them, but a lot of them—who saw a potential disaster on the horizon. Burning fossil fuels emits carbon dioxide; what if global warming was the result of the rapid increase in fossil fuel consumption? The group scampered back to the computers (this was in about the year 10 BG, Before Google) and correlated the rise in temperature to the rise in fossil fuels. Lo and behold, they appeared to move in lockstep, and a new earth-shaking baby was borne unto the world: the news that our reliance on fossil fuels may well be dooming the planet. This led to a global movement to stop usage of fossil fuels.

With that, and metaphorically speaking, two massive and unfriendly cats were dumped into the same bag, and here we are.

This quasi-analysis skips over a lot of important factors that would be significant to consider in this fossil fuel/rising temperature correlation: the rise of nuclear power, the increase in natural gas usage, the continued reliance on coal, and more. They will be discussed later. Proponents of fossil fuels see a world absolutely reliant on their use, a population that would die without fossil fuels or, even worse, be forced to give up all creature comforts. In the short term, they are right. In the medium term, we don't know if a total conversion from fossil fuels is even possible, but for some people, the question is too ludicrous to even contemplate.

Proponents of climate change see a world in grave danger, one that can only be saved if we get off fossil fuels at warp speed and figure it out as we go along.

You probably have a burning question on your mind: who is right? I'll not beat around the bush; I have no freaking clue. And that's fine, because we shouldn't even be considering that question. By framing the debate that way, we force people to consider it as a binary alternative: either support the environment, or support fossil fuels.

Most people who pay attention either to energy or the environment have fallen into one of these two ruts, and they can't get out. The good news is that it is possible to view this debate in an entirely different way that can reframe it so that both sides stop wasting energy fighting each other and start finding solutions. But before getting to that, we need to look at a few of the facets a little more closely, and then the answer will pop out like a dynamited gopher.

There is some bad news too though. It's true that people who are concerned about these things tend to fall into one of the two ruts mentioned above, but what's sad and disturbing is that the vast majority of people simply couldn't care less. Oh, everyone *cares* about the environment, just like everyone cares about the dying refugee we see on TV. And everyone cares about whether they have any fuel to heat their home or fill their vehicle or populate the grocery store, just as they care about whether the sun will come up tomorrow. Hopefully, you are concerned about either the environment or the fossil fuel industry, and if not concerned at least curious; for all your acquaintances that don't really care about either, they should. Please send them copies of this book (discounts available for orders of 100 copies or more).

5.
Environmentalists are right – oil usage trajectory is unsustainable

ROAD TRIP! ROAD TRIP!

What does the phrase "road trip" mean to you? Hours and hours on the road, alone or with buddies or family, stops every hour or so for junk food/gas/pee breaks, venturing off to another city or to see a concert or family or in search of an experience. A memorable philosophy professor I once had used to climb in a car with his buddies for weekends; each would write a bunch of ideas on pieces of paper and put them all in a hat, and they'd randomly draw a half dozen, and there was their weekend. (To be fair, he said that on some weekends, they never left the driveway, but he was also the first person I'd met who extolled the virtues of acid, so there you go.) For the rest of us who yearn for adventure in bite-size pieces (or ones that aren't quite so weird), there are countless websites documenting "drives to do before you die" or similar jaw-dropping must-see journeys that are guaranteed to make your local escapades seem crappy and inadequate. Sites like the Lonely Planet offer up Six epic drives of the world,[1] and simply googling "road trips" brings up over 51 million ideas to sort through before heading off again.

Now who do we suppose is doing all this road tripping? We might think everyone is, and that's ethnocentrically true, if what you mean by "everyone" is "people like me." That's how people tend to view the world. But from a global perspective, there is some substantial value in refining exactly what that means.

Road tripping is primarily a middle-class North American or European phenomenon, because we have the means, the cars, and the interest to do so. The wealthy of the world, no matter where they live, don't really go on road trips; one seldom stumbles across the children of Mexican billionaires at a truck stop off the interstate. The poor of the world don't road trip because they're, well, poor. No, it's a cultural phenomenon based on the desires and abilities of the middle class of the western world.

We also now dance around a pot called our bucket lists. Not everyone has them, but many do, and they tend to be glorious and somewhat outlandish. They are that way because that's the point—they are there to push us to extremes, outside of our comfort zone, and to ensure that we experience all of this globe that we can before we are planted into it. For more adventurous types, these lists include climbing mountains on all continents, or visiting Antarctica and Bangladesh in the same week, or other randomly bombastic stuff that will *kill* on social media. Few people's bucket lists include things like "go to the library" or "venture into the next city for a weekend." Bucket lists are homages to covering the world in as extreme a manner as possible (or just slightly outside of comfortable, but not very far) for fear of missing out.

With regards to fossil fuels, consider these recreational pursuits from two vantage points, the first being before we could afford much leisure. What would have been on a bucket list a hundred years ago? Or two hundred? For almost all of the world's population, the words "get food" and "get firewood" would have featured prominently, as without these items, they would have been kicking that bucket within a week. Fuel was not consumed for casual gallivanting in those days.

Now, how about a second vantage point, that of the rest of the world? What about the youth of Mumbai or Mexico City, the combined population of which equals probably half the US road trippers' demographic? What if they get their mitts on a car for the first time and decide they want to see the six epic drives of the world, from Namibia to Australia to the US to Switzerland to places where there are not even any roads? What if those parts of the world that are just mass-discovering the automobile happen to have 4 billion people?

This brings us to an interesting point, probably the most interesting one in the whole world of climate wars that no one even realizes: regardless of whether you drill oil wells for a living or are a Greenpeace warrior, the realization is the same that if the world adapts to the North American/European way of life, the stress on resources and energy will be far more than substantial. The scary part of that aspect is the trajectory. If the whole world starts road tripping like we do, global energy requirements will reach epic proportions.

WHY IS THIS A PROBLEM FOR BOTH SIDES OF THE FOSSIL FUEL DEBATE?

We can easily see why opponents of fossil fuels fear the coming of a new global generation of middle class travellers. Climate warriors should worry about mass adoption of the travel bug; they have a full-frontal assault on fossil fuels and automobiles as it is now, and a billion more travellers will inevitably mean more fossil fuel consumption. Green pressure has led many governments to hand out mind-boggling subsidies to consumers just to buy green-ish vehicles. These hybrid vehicles, which run on both electricity and gasoline, often attract subsidies simply for being more efficient, which is perplexing because bigger benefits would be gained by incentivizing light truck and SUV drivers to simply downsize into something more efficient. No one ever said government policy had to make sense. In other jurisdictions, a certain percentage of vehicles will soon need to be green or all electric. At any rate, it is clear why the environmental movement should not want to see a mass migration from non-travellers to junk-food and caffeine-fueled road warriors.

On the other hand, you might think that the petroleum industry would be overjoyed if that were to happen. Several billion people adopting a drive-anywhere mentality would be good for the fossil fuel business. That is definitely true in the short term; sharply rising demand would no doubt generate a lot of cash flow for a few years, and a lot of the industry would see nothing but glory days ahead.

But there would be trouble for them before too long as well. Several issues will be the cement pillar to their speeding sports car. A sociological one is that the environmental movement is doing a phenomenal job of turning the world against oil companies. Investment funds are being persuaded to dump energy shares, and infrastructure development is being thwarted at every turn by some young idealist touching a piece of equipment for the first (and probably last) time in their lives when they chain themselves to it in protest of some construction project. Development is therefore becoming much harder. Whereas once big oil roamed the earth like unstoppable dinosaurs, probing their proboscis into the earth here and there in the search for new deposits, that game is now largely over. Petroleum exploration is about as welcome globally as a guy sitting outside an elementary school in a dirty van with no windows.

It's also getting harder because the world is close to being picked over. If you pay attention to world oil discoveries in the news, you will find that there aren't any—not of any size, anyway—and those that are found are in the most remote and inhospitable places imaginable. New discoveries are almost overwhelmingly hundreds of miles from any civilization or off some exotic shore, or in Angola or such locations that seldom make it onto bucket lists. You may have read a lot about the US shale revolution, a development that has absolutely captivated the world's media. These deposits are indeed large, but after a decade and probably close to a trillion dollars of investment, they have increased global output by about five percent. Six if you're an optimist. Like all deposits, they too are finite.

The world's petroleum industry will therefore be hard pressed to maintain production levels beyond the next 20-30 years, because at today's consumption level, which is growing steadily, the easy-to-come-by stuff is going to be long gone. Yes, there is a lot of oil in the ground, but that doesn't mean it is economic to produce easily. Some is, some isn't. There is also a lot of gold in the ground, but would you grind up a mountain to get three ounces of it? The same goes for much of the world's oil; it will take vastly higher prices for a lot of the world's reserves to be developed.

What if demand actually accelerates, or even keeps growing? To environmental groups, that scenario is like sunshine to a vampire, but that doesn't mean it won't happen.

WAIT A MINUTE... HOW MANY CARS ARE WE TALKING ABOUT?

There are about 2 billion working cars and light trucks on the planet. You could view that as one light vehicle for every 3.5 people, or, as the industry

ENVIRONMENTALISTS ARE RIGHT – OIL USAGE TRAJECTORY IS UNSUSTAINABLE

prefers to quantify it, something like 300 cars per thousand (I will refer to light vehicles as cars for brevity).

To get a feel for the magnitude of these numbers, it is far more illuminating to compare regions. Have a look at 2014 data[2] for a handful of regions and how many cars are owned in each.

2014 stats (cars per thousand people):

Africa:	35	(trend is flat over last 10 years)
India:	31	(increased 3x in 10 years)
China:	104	(increased 5x in 10 years)
USA:	815	(flat to slightly lower in last 10 years)

Now it's getting interesting, or terrifying, depending on your perspective. Note how India's and China's rates are increasing. These numbers are a reflection of their growing economies and each country's ability to raise its standard of living, which seems inevitably to turn into auto sales.

Now suppose those regions of the world keep growing their economies and standards of living until they approximate those of the US. On second thought, forget that; several billion people matching the living standard of the US would denude the entire planet in weeks. The US has an untouchable lead in hyper-consumerism. For China and India to get to that level is pretty much impossible, or we should hope anyway. So, let's say they get halfway there, and we'll assume that vehicular ownership is a reflection of this newfound level of consumerism.

We could then apply comparable vehicle ownership statistics to these considerable populations:

Africa:	1.2 billion
India:	1.3 billion
China:	1.4 billion

And for reference, remember:

| USA: | 0.3 billion |

You can see where this is going. If India and China reached one half of the US' vehicle ownership per capita stats, China would add 425 million cars, and India would add 490 million cars.

Now that's just cars, and that's just crazy. Those totals, if they happened (and that's assuming that Africa doesn't catch up to those levels, being quite far behind at this point), would increase the world's car population by more than 40 percent. Bah, you might say, don't be crazy, that's not going to happen; the world is heading the other way and we're all about being green now. Except that it's not crazy at all, it actually is happening. China is expecting annual vehicle sales to rise to 35 million per year by 2025,[3] which would be more consistent with a hypothesis that they will add 425 million vehicles than with the idea that green energy is taking over the country, or the previously discussed idea that petrol cars will no longer be available in eight years.

But wait, you say, you're forgetting electric vehicles. Am I?

WHAT ABOUT ELECTRIC VEHICLES?

At this stage in history, a perusal of mainstream media would convince you that electric vehicles (EVs) have all but taken over. So, isn't it a bit stupid to conclude that vehicle sales growth in developing countries will create soaring demand for petroleum?

First, let's dissect the current conventional wisdom. The news has more coverage of EVs than of Donald Trump, and the fact that they're even comparable gives you an example of the state of EV madness out there. The only thing comparable to Trump's impact on the globe would be a volcano that covers Europe and Asia with lava. But the press loves the topic, and the auto industry knows a trend when it sees one (the fact that governments have guns to their heads to ensure EVs are flooding the market has something to do with it too, but still…). Major manufacturers are stumbling over themselves to announce new models, or, like Volvo, are creating headlines implying that their entire range will be electric by 2019[*]. Additionally, EV

---

[*] This bit of marketing genius was in fact quite disingenuous; Volvo's actual 2017 announcement was that all its products would be "electrified," as in either pure EV or a with hybrid gasoline/electric powertrain. This is a critical distinction, because hybrids can offer EV-type fuel efficiency if driven with butterfly-weight pressure on the go pedal,

ENVIRONMENTALISTS ARE RIGHT – OIL USAGE TRAJECTORY IS UNSUSTAINABLE

owners are wildly enthusiastic. Tesla owners love their Teslas, and say so a lot. I love their Teslas too, but the world is more complicated than that.

The EV phenomenon is the environmental movement's tactical trump card, the avenue by which this elephantine problem of a soaring vehicle fleet can be managed without destroying the environment. The climate movement has a two-pronged strategy to reduce fossil fuel usage for transportation: attack oil producers and block further oil production, and promote the adoption of EVs. Attacking oil production clearly isn't working, primarily because consumption continues to grow year over year and this demand ensures new supplies are found. The tactic of strangling supply could theoretically work over time with enough effort, but it doesn't appear to be on track to meet environmentalists' goals of halving fossil fuel consumption by 2050. Few forecasts see consumption diminishing by that amount, a position supported by soaring global SUV sales. Full-size pickup trucks are also rising in popularity; they are the number one selling vehicles in North America, and they seem to get larger and more luxurious every year, with corresponding sales increases.

EVs' capabilities are definitely becoming impressive, with ever-longer ranges, more compact batteries, and a growing raft of choices available to consumers. Charging stations are sprouting everywhere to accommodate travel plans. The wheels of the EV revolution are in motion, no doubt largely due to the heavy hammer of government, of which the world's more progressive (and debt-loving) examples are pouring on subsidies to purchase EVs and subsidizing construction of charging networks.

But the EV phenomenon brings several continent-sized problems to light. The largest relates to the system discussed earlier: the global energy generation and distribution system that we drive over, under, and beside every day but never notice at all.

Based on headlines, EVs appear to be taking over, but that is an illusion created by the tiny starting point. This is what legitimizes the countless headlines bleating, "EV sales are soaring." Sales really do soar in percentage terms when they go from five thousand to ten thousand,

---

and for short distances, or can consume fuel with the best of them if driven the way a lot of people do. Volvo cleverly declined to correct the wave of incorrect media reporting of the announcement.

which garners a headline, but are utterly insignificant when compared to growth in Chinese non-EV sales that increase by two percent, which is over three hundred thousand cars. What will happen to the system if EV sales became non-miniscule?

That's where the wheels fall off, because for EVs to take over, or even get close, the system would need a mega-rewiring that any electrical engineer would lose his mind contemplating. Highway EV charging stations work well enough when one in every thousand cars is an EV; what if it were one in two? This is but one tiny example, so let's consider another simple one: modern suburban or apartment-complex electrical systems. A lot of specific hardware brings power from major transmission lines down through substations, into lower voltage distribution systems, and finally to relatively low-voltage neighbourhoods. What if a majority of the vehicles in those neighbourhoods became EVs, and home fast-chargers were installed everywhere? What would be required to rewire the grids to handle this? And who would pay for it?

Proponents point out that solar installations and large home batteries are on the way, and not only will these provide home-grown power but that the excess can be sold back into the grid. And the cost of doing that for an industrialized city would be? Hmm, it's gotten deathly quiet.

One of the reasons is that, in the current day and age, it is foolhardy to predict the cost of the major systemic overhauls that would be required. Infrastructure is no longer just planned and built; the process now includes consulting anyone who could theoretically be impacted, politely listening to the reasonable and not at all reasonable suggestions and diatribes, and then some years later cobbling together the necessary licenses to proceed. That is the process for any single project. Now imagine rewiring an entire city or country. Those who say it is not difficult have clearly not built anything bigger than a dog house during this century.

Environmental activists may dismiss the above as a lazy analysis that isn't reflective of reality. While it is hard to argue about the laziness, the last word can be left with one the fastest-growing and staggeringly huge jurisdictions mentioned above. The Chinese government is forecasting 35 million vehicle sales per year by 2025 (the same year Tony Seba predicts that petroleum powered cars will no longer be sold), and it is expecting only 7 million of those to be electric. China is therefore still on track to be adding some 400 million petroleum powered cars in the next fifteen years. Add in India, Africa, and whoever else that has a rapidly growing economy and prefers the California lifestyle to the Mumbai one, and it is unlikely that petroleum usage will be halved, dented, or even meaningfully reduced in the next decade. It certainly won't be eliminated by 2050, the year picked as a key turning point in the global temperature change debate. We will be lucky if the world's consumption is down to 75 million b/d by then.

Studies turn up periodically, such as one by a bunch of academics from a Dutch university,[4] that indicate that fossil fuel usage will be pretty much done for by 2050. Intense statistical analysis ensues that extends historical trends to conclude that losers will be, as they put it, "Russia, the United States, or Canada, which could see their fossil fuel industries nearly shut down" because demand is going to utterly disappear. How this will happen is explained in mind-warping circular references to the authors' own models, which appear to bear no tangible reference to reality. For example, the authors dismiss current modeling efforts and declare the superiority of their own with this sweeping comment: "Such models struggle to represent the effects of imperfect information and foresight for real-world agents and investors. By contrast, a dynamic simulation-based model relying on empirical data on socio-economic and technology diffusion trajectories can better serve this purpose." To an academic, that might sound like a properly rational statement, and one can see them gathering around their model and enthusiastically nodding at the brilliant output that shows fossil fuel usage falling to zero. The only problem is that the "empirical data" on which it is based is mysteriously undefined, because the real empirical data, such as the fact that fossil fuel usage is currently climbing by over 1 million barrels per day annually, is at least somewhat pertinent, to put it mildly. More realistically is the type of empirical data that their model carefully circumvents. The whole report is a litany of inconceivable suggestions, such as that solar power will replace natural gas, which will be fascinating news to those who live in northern climates where the sun is hardly seen in winter.

If the conclusions of such studies are too difficult to understand, there is a reason for that, and it's not because they are PhDs and you are not; it's

THE END OF FOSSIL FUEL INSANITY

because they have an irrational faith in their models and will defend them long past the point where they head off in a different direction than reality.

That wacko report did serve its purpose, though, by finding its way into the mainstream media and proving once again how dangerous it is to simply follow the headlines in the popular press. Canada's Global News took the bait and treated it as useful information,[5] commenting: "Canada's oil sands industry would be among the first to collapse in a possible global financial crisis within the next 15-20 years, which is when a new study predicts clean energy-technologies will all but kill demand for fossil fuels worldwide." It is no coincidence that Global News came up with this snappy sound bite by interviewing the author; there is not a chance in a billion years that they would have come to the conclusion that the author did in interpreting his own thesis. The author started with a conclusion, built a model that validated it, and went on the news to "fill in the blanks" for an uncritical media analysis.

This is why we are here, pondering these things in this book. We are here because such ridiculous headlines for ridiculous reports are what any interested citizen is faced with when trying to find energy answers in mainstream media.

NO ONE IS GOING TO BE HAPPY

That is correct. As cheap, easy-to-extract oil reserves are depleted and consumption continues to rise, the petroleum industry is going to spend a decade or two elbows deep in cash, but at the same time, a paucity of new global resources will mean that their growth model will have a finite shelf life. The world has literally been picked over, and the industry will be extremely hard pressed to keep up with demand in an environment where the entire business is being treated like a mass murderer (not my words; Arnold Schwarzenegger publicly accuses big oil of mass murder,[6] and in contrast to his acting career, I have not read any negative reviews of that performance).

Environmentalists are going to be as wild as pepper-sprayed bears if oil consumption continues to grow, but it looks likely that it will. The numbers are simply too overwhelming. If either India or China reaches half the auto ownership density of the US, and any decent portion of those vehicles uses

ENVIRONMENTALISTS ARE RIGHT – OIL USAGE TRAJECTORY IS UNSUSTAINABLE

fossil fuels (as most are likely to), global oil consumption will increase by tens of millions of barrels per day.

This is the point where both environmentalists and the oil industry should be getting pale—the environmental industry for obvious reasons, and the petroleum industry because the cookie jar is getting empty. The oil industry would have extreme difficulty responding to that sort of demand increase. Oil industry advocates, superficial energy pundits in the media, and well-paid securities analysts point to the US shale revolution as being capable of meeting this demand. But even a cursory, high-level look at that data shows it can't come close. The US shale revolution has barely replaced global natural production declines over the past decade, at a cost of probably a trillion dollars, and in the process, has devoured much of the sweet spots of shale deposits. Shale growth will certainly continue and add a few more million barrels per day, but tens of millions? Not a chance.

THE ABSOLUTE VERSUS PER CAPITA DEBATE

One school of thought considers it unacceptable and overly hypocritical of westerners to "blame" China and India for climate change issues. This line of thinking (as outlined by Hans Rosling in his book *Factfulness*[7]) goes something like this: China and India have a lot of people, but generate few emissions per capita; the west has few people but generates vastly more emissions per capita. Therefore, it is inhumane to deny developing countries our standard of living. Westerners are the problem because of the high per capita emissions.

That line of thinking is probably a useful pillar in sociological studies, but is violently disconnected from the raging fossil fuel debate. Climate activists are rather black and white on the issue: CO2 levels are unacceptably high and cannot rise much further. Therefore, fossil fuel usage must stop. There is no room in that argument for the contemplation of relative contributions. If their argument is to be believed, then absolute contributions are all that matter.

From that climate change perspective—that absolute levels of CO2 must be reduced—if one wants to talk about per capita contributions, the only path forward is to talk about the elephant in the room: that there are too many people on the earth. Good luck with that one. I'll leave it to someone else to work that out. I will leave the per-capita vs. absolute emissions levels discussion with these two observations: Piers Forster, director of the Priestley International Centre for Climate at the University of Leeds, commented in a 2017 Guardian article[8] about how global carbon dioxide levels had surpassed 400 parts per million. He writes, "These large increase [sic] show it is more important than ever to reduce our emissions to zero—and as soon as possible…" In a 2017 Reuters news article[9] about India's plans to build more coal-fired power plants, it was noted that "Carbon dioxide emissions from India's thermal plants are expected to jump to 1,165 million tonnes by 2026/27 from 462 million tonnes in 2005, the CEA estimates. Emission intensity, measured in carbon dioxide emissions versus GDP, is likely to fall, however." From this, we can see that falling emissions per unit of measure—whether GDP or person—is useful from certain sociological perspectives, but is of no help to global CO_2 parts per million. If that is what we truly need to be worried about, the concentration of CO_2 in the atmosphere, it doesn't matter if all the emissions in the world come from a single Norwegian or ten billion goat farmers.

6.
The electrical utility guy is also right though. Wait, who's that again?

"When we generate electricity from a conventional generator, be it coal, gas, oil, or hydro, the machines are all linked together by the transmission system. This synchronizes all of the generators, which in North America is at a speed of 60HZ or 60 cycles per second. To put it simply and without going into great detail, the magnetic force of all of these generators in synchronism gives the system stability, both steady state and transient, which keeps the whole system operating in a stable state and able to withstand line trips, generator trips or loss of load without taking the entire system down by loss of angular stability or a cascading voltage collapse. The more of these independent power producers generate back into the grid, and the more we depend on them for energy to feed load, the less stable our system becomes. With all of these energy sources on during the night that do not offer any spinning mass (inertia) to the system, the less stable the system becomes and therefore the less reliable."

—Peter Gibson, 40-plus years' experience as an electrical utility sector employee

THE END OF FOSSIL FUEL INSANITY

I'll level with you: I have no idea what this guy is talking about. I know energy fairly well, having seen many facets in the past few decades, but as far as this end of the business goes, I am as ignorant as trust fund kid in a laundromat.

Here is one thing I do know about electrical systems, an example that is more pragmatic and tangible. One fine summer day when I was young, a power outage impacted a sizeable region of my home province. Apparently, a cat had gotten into one of the larger central power substations, whereupon it appeared to complete a circuit of some sort and used up all its nine lives in a brief but no doubt spectacular blaze of glory. Several thousand square miles of agricultural and small-town life spent the rest of the day without power, and over one corner of it hung a haze of cat smoke.

I learned several lessons that day. First, never throw cats around near electrical power stations, but I guess I kind of knew that anyway. Second, the power grid is, despite its ubiquity and strength, in some ways as fragile as a kitten.

Most people, including the ones who want to rewire the power grid tomorrow to eliminate fossil fuels and replace them with renewables, have a level of understanding of the grid that is similar to my cat-frying knowledge base, as opposed to the expertise of Mr. Gibson quoted above. The number of people that understand electrical grids, how they work, and what stresses them, is probably limited to the number of people who are paid to know that stuff.

Big deal, you might think. No one except a handful of mega-geeks knows how to make a computer chip either and we seem to be getting along just fine. That part is definitely true, as we saw earlier. There are all sorts of systems that the average citizen doesn't even know exist, that we rely on for almost everything in our lives, and we are clueless about those too. The world as we know it keeps turning.

There is one humungous difference though. We have an electrical system that works well, that was built over the past hundred years. What is so different about the electrical system and our poor knowledge of it is how we are, on a global basis, running full speed ahead towards a new electrical power-based world powered by renewable energy, and that world is shockingly incompatible with the one we have now. The total number of

THE ELECTRICAL UTILITY GUY IS ALSO RIGHT THOUGH. WAIT, WHO'S THAT AGAIN?

people in the world that either understand all this or care to would fit in a couple of garbage dumpsters.

DANGER, HIGH VOLTAGE – THIS ISN'T A SEWER PIPE YOU'RE MESSING WITH

Previously, we stopped in for a visit at the Calgary traffic planning department, a cauldron of frustration, fear, and hopelessness. Its denizens are kept out of a mental institution only through the voluminous use of tranquilizers and sedatives (as near as I can tell from their output). What has savaged their will to live and their ability to create anything sensible is the immutable law that whatever they do has to be consistent with existing infrastructure, and yet it has to meet new demands for which the old setup clearly is not capable.

Now, keep in mind that this is traffic we're talking about. Obtuse road planning and criminally stupid traffic solutions are, at the end of the day, just irritants (possibly are even tourist attractions, as when a simple right turn becomes a sequence of traffic circles and overpasses spanning forty acres that renders a car's navigation system senseless). The point is that there is much difficulty in adapting an existing system to meet new loads or growth.

Now imagine what that would be like with an electrical system. This isn't going to just be a lecture about high-voltage transformers, copper wire, and dead cats; we don't need to get into those details to understand the challenge at hand. What we need to look at briefly is what happens when governments, organizations, special interest groups, and a whole new industry are encouraged to develop renewable energy sources and ram the output into a system they don't understand either, and expect it all to work.

WORLD'S WORST ELCTRICAL GRID OVERVIEW, BUT IT'LL DO

I understand quite well that I'm the last person that should be guiding you through the inner workings of the world's electrical system. I have started electrical fires when making breakfast and have installed light fixtures whose switches subsequently work in theoretically impossible ways. An electrical engineer could do a far better job of describing the electrical grid, but you wouldn't notice because he/she would be far too clever and full of dreary

43

facts and you wouldn't be listening. We do need to check in with them now and again, as with the quote above from the utilities fellow, but only to know that the waters are far deeper and more mysterious than we would otherwise ever know.

This ignorance actually proves a point in some ways. People at my level of obtuseness are now, by government mandate, able to install wind turbines or solar panels and wire them directly into the grid to sell home-made power.

The whole world is rushing headlong into an electrified future. We are told that gasoline will soon be obsolete or forcibly kicked to the curb, and not just by environmentalists. Countries like France and China have mused out loud about plans to eliminate internal combustion engines within the next few decades, which will be replaced with EVs. We also hear that, for example, California has mandated solar panels on rooftops of all new residential construction, and that power has to go somewhere. It is therefore evident that the electrical power system is expected to handle the burden of transitioning to a greener future.

This raises a few questions: Does anyone know what that all means and entails, and who is planning for it? Why do we make the assumption that power can be treated this way, that almost any new power production is acceptable into a system not designed for it, when no one appears to have even considered what this would mean in terms of adapting the power grid?

As it turns out, a few people have considered this, and their findings are eye-opening.

THE ELECTRICAL GRID 101

As promised, this examination won't get into the minutiae of power delivery. What is relevant to the topic at hand is how this is yet another massively entrenched system that will be beyond difficult to modify. To rewire the electrical grid represents a challenge similar to what, say, Germany would face if required to govern using Italian standards.

The electrical grid system was largely developed about a hundred years ago. Like the traffic systems built in conjunction with the rise of the automobile, it was right for the time. Power was generated in specific locations, such as where coal fired power plants were constructed or hydroelectric dams built, and large transmission lines carried power off to consumers.

THE ELECTRICAL UTILITY GUY IS ALSO RIGHT THOUGH. WAIT, WHO'S THAT AGAIN?

Eventually, nuclear power was added here and there, and flexible natural gas-burning power stations also. Small power sources made little sense because of the capital outlay required to move that power—quantity mattered, like most industrial products. No electrical utility would spend the considerable amount required to connect some inventor's home-made windmill into the grid, because the time required to recoup the investment would have been thousands of years.

A system thereby developed around these large-scale power sources, a distribution grid that could reliably coordinate and disseminate the electricity to wherever pockets of demand were materializing. Over time, the various power sources bumped into each other and became interconnected, thereby forming the massive power grid as we know it. These interconnections were extremely valuable, as a particular residential grid no longer had to rely on a single source of power. It could come from any number of different sources that were connected to the grid. But the sources of that electricity were a finite number of large power generating installations, and the web of transmission lines reflected that structure.

Big power sources generate electricity in such large quantities that it can only be transported by wire in those huge high voltage systems you see here and there, the big towers with the power lines so high up that copper thieves don't even try. These main lines carry power to substations, where the voltage is reduced from the massive levels that can flash-fry a cat in a millisecond down to the level that you can experience in your house, which is merely unpleasant to experience in person but can be quite funny when experienced vicariously on YouTube. This system is and has been the only way to move the extremely large quantities of power from source destinations and distribute it in the low-voltage form that suits our appliances and allows us to mishandle electricity without encountering the same fate as the power station cat.

All is well and good then; the major power sources are all interconnected and contribute their high voltage input into the system, where it is then carried to major substation nodes, where it is then weakened to a lower voltage, and then distributed via low voltage systems to pretty much everywhere. The system can also handle a critical variable, the fact that electrical power consumption fluctuates wildly during the day. It is far higher at certain times of the day, like the dinner hour, when everyone returns home after their day out and about, and turns on every conceivable device in the house. Contrast this with the power load at 3 am, which is negligible and almost non-existent—with the exception of fifteen-year-old video gamers who hide out in the basement. All in all, the electrical system functions rather well as configured.

Then, here comes trouble.

The problem is best illustrated by way of analogy. Imagine you had a fleet of buses used to haul children to schools across the school district. You have just enough of them that the buses are pleasantly full but not overcrowded. Then one day a politician or traffic planner notes that garbage trucks run around empty most of the day, and so do school buses. To a politician or traffic planner, there is a brilliant solution: drop off the kids, then circulate the buses to pick up the trash.

There's nothing fundamentally impossible about that idea (I'm not talking practically; I'm talking theoretically). In a way, that is what we're facing now.

That's the state that the headlong rush for green energy has brought us to. We are being told to use a system designed to move kids to now move garbage. Now, before howling in outrage, this is an imprecise analogy and possibly demeaning to green energy advocates: if you prefer to switch it around and not refer to green energy as garbage, let's assume that the program was reversed, to haul all the children in garbage trucks. Now we can all calm down.

The energy grid is designed to move energy in one direction: from major power sources through ever smaller "pipes," right to your electric toothbrush. It has meters along the way so we know what's happening, and it has all that operationally complicated stuff Mr. Gibson mentioned earlier.

The power grid is a major part of the energy system that makes our world work the way we want. Rewiring it to handle a purely electrical future devoid of fossil fuels is like rebuilding the US interstate system by moving the entire thing exactly a hundred meters to the left. The professionals who run these systems don't yet understand the full ramifications of trying to adapt the grid, yet there are countless websites with endless stories about how the "smart grid" and "micro grids" and small-scale storage/withdrawal

THE ELECTRICAL UTILITY GUY IS ALSO RIGHT THOUGH. WAIT, WHO'S THAT AGAIN?

schemes are going to change the world. It sounds wonderful in theory, if one assumes the grid is readily adaptable to these changes. It is not. Here is one example of how difficult it would be, pulled from the pile of many, presented for no other reason than I happened to learn about it from a friend who is in the business.

IEC 61850 EXAMPLE – A CRITICAL, UNKNOWN STANDARD THAT SMART-GRIDS REQUIRE

IEC 61850 is a relatively new international standard that almost no one has heard of, and possibly, you never will again after this chapter. But that doesn't mean it's insignificant; it is in actuality relevant beyond belief, and you should consider that the next time you're thinking of something to Google and you go for the Kardashians instead. It really does have a fascinating story behind it, if your level of interest rises above the tabloid swamp. I profoundly hope so, for your sake.

According to Wikipedia[1], IEC 61850 is "an international standard defining communication protocols for intelligent electronic devices at electrical substations." Another website gives a useful analogy: "Imagine having hundreds of manufacturing plants scattered across the country, all with their own brands of devices communicating in a wide spectrum of protocols. Now imagine that in order for these plants to function efficiently and effectively, there is a need for these plants to communicate with each other BUT the protocols are not interoperable; the devices are not meant to communicate with each other natively. Replace these manufacturing plants with substations and you have an idea what is happening throughout the world with regards to power transmission and distribution. In order to maintain guaranteed delivery times, power quality and reliability, substations need to exchange critical information amongst each other and with data centres. However, with the many protocols, much engineering work is required."[2] In the article, "Much engineering work" isn't defined, but could safely be interpreted as "holy sh_t, that's next to impossible." The article then points out how renewable energy and micro-grids introduce a whole new level of protocols and power characteristics that are capable of destabilizing the grid. The IEC 61850 standard defines communication profiles and data models that allow all these connections to communicate efficiently at substations and provide stability to the grid (which, as you'll recall, Mr. Gibson indicated is critical).

Two fascinating aspects of this protocol stand out to the average layperson. The first is that, in order for the world to be able to go to a truly "smart grid," where small connections can remove or inject power into the system, all need to be able to communicate properly and incredibly quickly—millisecond quickly—to facilitate grid stability.

Without the ability to enable and control this new stratum of activity, there will be no smart grids and no utopia of micro-generating sites contributing to the grid. And now comes the second, really interesting part.

A handful of dominant players in the energy infrastructure business built a lot of the world's main grid systems a hundred years ago. They also operate and maintain them. Now consider this: none of the big players know how to entirely implement this new standard. None of them. Even worse, as one mystified and slightly indignant expert put it, "It seems that none of the big players truly understand digital communication, otherwise why would the key profiles defined under this standard be based on a full seven layers of the academic-based OSI reference model? Where else but in schools for academic purposes is the full 7-OSI model being used? The internet, as the largest successful human-built communication network, definitely is not an example." Don't worry, he lost me too. What's significant about these comments though is that it is an indication of the hurdles that even the system's architects will face in evolving the system.

The expert noted above is Damir Karagic, one of the kingpins in the fledgling IEC 61850 business, and the founder and owner of a company called Grid Systems. He has his hands full. The world's biggest companies beat a path to *his* door, asking for his help.

Mr. Karagic's company is one of the few entities in the world to have been certified to the full IEC 61850 standard, and his shop has fewer than twenty people. The big players, the ones who built and control most of the world's substation infrastructure, have not even caught up with the global standard adopted *prior* to IEC 61850. The task to bring every substation in, say, North America is daunting beyond belief. A 2010 study[3] pegged required system upgrades at $1.5 to $2 trillion, and that was before even contemplating this new standard. It was also eight years ago, and the world has gone smart-grid crazy since. Accelerating those upgrades, if they can be done at all, will simply drive the cost higher. Mr. Karagic estimates the cost of upgrading a substation to IEC 61850 to be $1 million. *Each*. In the US alone, there are 70,000 substations, which implies a total cost of $70 billion. And that is simply so they can talk to each other.

Mr. Karagic, however, was just getting wound up. He went on to elaborate on how shockingly poor the current state of the grid really is. "Real-time operation of a power system has always depended on communication. The first real-time automation system at the beginning of the twentieth century (1930s) was based on telephone switching. Even though we live in the era of 100 GBS (gigabits per second) links, today's average automation data paths are at a pathetic 64 kbs (kilobits per second), some 1.5 million times slower. That's not bad considering the majority of power system equipment in North America cannot communicate at all."

I'm exhausted now, wandering around on unfamiliar terrain, but you get the picture. Well, you aren't getting the whole picture; IEC 61850 is only one small part of it. Adoption of this new standard will only allow substations to communicate, and there is no mention yet of what would be required to rewire the electrical system to bypass fossil fuels or to add smart-grid connections. The task seems so overwhelming that it introduces fits of giggles and light-headedness like one gets after being awake for seventy-two hours straight.

BEWARE OF LOONY EV PREDICTIONS

There are some staggering predictions floating out there about how many EVs are being sold, the tidal wave of them coming from manufacturers, and the desire of some cities and countries to ban internal combustion engines. Combine these factors with the Apple-like enthusiasm for Tesla's EVs—people lined up to place *orders* for them, with deliveries a year out at least—and one might reasonably conclude that it's game over for gasoline.

That is an unlikely scenario for a number of reasons, such as that consumers simply aren't all that interested in EVs (despite years of promotions, government rebates, and the Tesla phenomenon, the gains in sales of gasoline SUVs far exceeds the gains in EVs). But in keeping with the theme of this chapter, let's have a quick look at what mass EV adoption would mean for that poor old electrical grid.

It would mean chaos. There is no getting around that, particularly if quick-charging stations were to become widespread. And of course they would. Who would want to wait twelve hours for a battery to charge?

Recall how the electrical grid gets electricity to your house by reducing the huge flows through ever-smaller wires until it gets to the relatively miniscule consumption of your house. Even with a few teenagers and 182 electronic devices, the average house's consumption is not that large.

Now imagine how that consumption would change if one or two fast chargers were added to every house. Imagine what that would mean for a city block of houses, and the little wires that bring power to that street. Would they need upgrading? Who knows? The point is that no one knows, because no one's thought about it, or at least the people pushing EVs haven't thought about it. How would an apartment building function if half (or more) of the tenants opted to utilize fast chargers? We know what happens when a single circuit gets overloaded, such as when a poorly-wired kitchen has a few appliances on the same breaker and it trips. The same problem would befall apartment buildings or city streets, except they don't have similar circuit breakers. Thus, we need to understand the consequences of such electrical-consumption remapping before we get too far down the road.

There are countless other infrastructure aspects to consider that could fill a book on their own. What's important to remember when reading thrilling green-energy-revolution predictions is to do an adequate level of research on the topic of how this will all come together. Unfortunately, that library is hideously empty.

MEGA BATTERIES COULD CHANGE EVERYTHING—BUT WE'VE BEEN SAYING THAT FOR A HUNDRED YEARS

Wind and solar energy are, without a doubt, capable of producing vast quantities of power. These sources are doing so now, thanks to the hundreds of billions that have been spent on them.

Before considering the impact of all that power, let me pose a question. If your athletic body requires 2,000 calories per day, how does it work out

THE ELECTRICAL UTILITY GUY IS ALSO RIGHT THOUGH. WAIT, WHO'S THAT AGAIN?

if you eat all 730,000 of those calories in the first week of January, and none for the rest of the year?

That, in a weird nutshell, is the problem with wind and solar energy. It's produced in vast quantities exactly when not needed, and there is not yet any functional way to store it until it is needed.

Power consumption peaks, as noted earlier, at certain times of the day. This phenomenon is as regular as clockwork. So is the fact that this peak does not coincide with when the sun's output is strongest, or when the wind blows the hardest.

These ill-timed resources can cause a lot of problems. The base load of the power systems, those that provide the bread and butter 24/7 supply, cannot be turned on or off at the drop a hat. We also don't know on any given day what the solar and wind output will be. Suppose the day turns out cloudless with a howling wind. Solar and wind power generation will soar, and more energy will be pumped into the grid than can be used for that part of the day, but there's no way to store it. Because power prices are set hourly, the impact can be bizarre. For example, in Germany in 2017, there were 146 hours when power prices went negative[4], and this has happened in other countries like Canada as well. That is, power producers had to pay someone to take their power. As a rule, paying someone to take your product is not a good business model.

If just one of the vast horde of engineers out there could come up with an efficient way to store this power, it would be a game changer. Current-technology batteries don't work on any significant scale, because they are too expensive and cumbersome and do not hold all that much power relative to, say, a wind farm's output. Plus, as with a EV, they take time to charge.

Intriguingly, a battery can actually take many forms, if one looks at it as simply an energy storage device. An anvil on top of a ladder is an energy storage device, as opposed to an anvil on the floor, and you can get an idea of the energy involved by letting it fall on your foot. To the curious mind, energy storage ideas abound; anywhere that natural forces like gravity can move something, there is the potential to harvest that energy. However, all these super simple energy storage solutions seem to get thwarted in the real world, so there must be some difficulties that are not readily apparent to the average rube. For example, I've stood on a beach in Nova Scotia, watching the massive tides come and go before my eyes every six hours or so (I have an unusually high threshold for boredom). To my simple brain, it seems obvious that a tank could be placed on the beach at low tide, with an opening in the side that will allow it to fill up. Then as the tide recedes, the water could exit the tank in a way that spins a generator, or the generator could be spun when the water enters the tank as well. I beam at my cleverness and look forward to announcing to humanity that I've got this whole renewable energy thing figured out. But then someone in the business points out that these ideas simply don't work for a number of reasons, and that it has been tried before. I don't know if he is sh_ting me or not, but probably not, because when I read the environmental news, the various attempts to store energy are infinitely more complex. Some are so complex that they appear just plain dumb. But I know they are not, because people are throwing serious money at them as opposed to watching the tides come in, which costs very little. Their bet is far bigger than mine, and it is logical to assume they've investigated all the cheaper alternatives that people like me can come up with. An example comes from the California desert, where engineers are trying to store excess solar heat by melting salt, which can then be solidified later in a reaction that eventually spins a turbine. My idea is pure genius and about one percent as complicated, if you ask me, but no one asks me, and they are spending a lot of money to melt salt and that hardly works either (otherwise, the world would be full of these contraptions).

It is sufficient to say then that solving the large-scale-battery energy storage problem is wildly challenging. That is, unless no one's thought of my tides-and-tank idea, in which case the napkin upon which all the engineering was done is for sale at a reasonable price.

An alternative does exist that would solve several problems: use the excess power to create hydrogen, which could then be used as a clean fuel in cars with fuel cells. The drawback to that concept is that the hydrogen infrastructure would cost billions. Some see that as a serious drawback, but we should have observed by now that doing anything at all to change our fueling system is going to have a staggering price tag, and maybe hydrogen's is no worse than any other. Other aspects to the hydrogen scheme will be discussed in a later chapter.

THE ELECTRICAL UTILITY GUY IS ALSO RIGHT THOUGH. WAIT, WHO'S THAT AGAIN?

We shall leave the topic of the electrical grid with a parting observation. The electrical grid, whose basic footprint was set in stone a hundred years ago, is vastly outdated technology. It is also so large and entrenched that uprooting it and starting over again, without unacceptable power disruptions, is for all intents and purposes impossible. Somewhere in between leaving it alone and rebuilding it, we need to find a path to making it work with what we have in mind with green energy. But we absolutely, critically, must do that homework before we can even consider a true green revolution.

Part of that homework is to have a look at how we got to where we are with fossil fuels, particularly oil. There was once a juncture point, at the nascent age of automobiles, where the jury was out as to what would be the best propellant—gasoline, steam, or electric power. For reasons unknown, history shows we headed off down the petroleum route, for better or worse. As when Microsoft's original crappy operating system became dominant because of the events of the day, so did dirty old oil achieve dominance. That path, and its repercussions, are worth having a quick look at as well, because the development of oil dominance, and how and why it happened, leads us directly to the quagmire we're in today.

7.
Legacy of big oil

A FEW CENTURIES OF COLONIAL HABITS KICKSTART THE GLOBAL OIL PHENOMENON

Legend has it that Manhattan was "acquired" from a local native band for twenty-four blankets or sixty Dutch guilders or something. I don't know if that's actually true, or what the currency was, but Google says it went down that way, and Google knows these things.

Whether it actually happened exactly like that is irrelevant to this discussion. What is relevant is that this transaction is such common knowledge that I, an ordinary goof from Canada, thinks the story is true or likely true, and that Google does also. This means that the whole world thinks so too.

That this obscure historical footnote is so widely familiar brings up a series of interesting questions relating to the impacts of that Manhattan transfer. An overarching question might be: How would you like to be the native that made that trade? Or, more pertinently today, how would you like to be one of the descendants of the native band that made that trade? Do you think it would be a little embarrassing to have the world continuously marvel at how that trade eventually panned out? Would the fact that the whole world considers it the worst trade in history have any impact on your present day bargaining positions?

That sequence of questions hints at an issue that we seldom think of, but that boils to the surface in unlikely places.

No one likes to become famous for making what in hindsight was a colossally bad decision. That poor native band, when this tale is recounted, comes out in one and only one way: they look like a bunch of idiots. The fact that the deal may not have seemed ridiculously wrong when it was made is never part of the legacy; all that comes to mind when we hear these bad-bargaining examples is "what were they thinking?" In the case of Manhattan, the heartburn to the seller's descendants would be profound, because everyone also knows (now) that Manhattan is the center of the universe and worth hundreds of billions. To be constantly reminded of this would make you want to throw up, every single day, and you would most certainly vow to never be shafted again.

The Manhattan example is perhaps the most visible and lopsided one around, but it is far from the only one. Colonialism littered history with dozens of similar "it was only driven on Sundays by an old lady" deals that have created cultural rifts that may never go away.

To delve into the consequences of these colonial hijinks would be loads of fun, but we'll resist the temptation. This chapter is about the legacy of Big Oil, and there is plenty to work with there too, for big oil has, if not quite as infamously as the Manhattan transaction, left its own fingerprint on the cultures of half the world. In fact, the history of big oil has played a huge role in the way our modern world looks.

BIG OIL

Big oil is the term used to describe the handful of petroleum supermajors, the multinational companies that form the global backbone of the oil industry. Once described as the Seven Sisters, these companies dominated the global petroleum scene from the 1940s to the early 1970s (and are still a formidably large component). During that period, the group controlled approximately 85 percent of the world's oil reserves. That is a shocking number for a handful of oil companies to control, and at some point, the countries that owned that oil decided so too, because those big oil descendants now control only about 6 percent. Based on that drop, you'd think those companies were a bunch of big numbskulls to have fumbled the ball as badly as the natives in the Manhattan transaction. Such is not the case, however; the astonishing drop in the share of reserves that they now own or control was a result of big oil doing similar trades as with the Manhattan natives, except that in the big oil instance the counterparties were able to eventually unwind the deals and regain control of their resources.

WHAT WAS SO BAD ABOUT BIG OIL?

Like many other historical events that look one way in hindsight, the question of whether actions were good or bad *at the time* is something else entirely.

Generally speaking, World War I (1914-18) marked the beginning of the petroleum age. Combatants entered the war on horse and ended it by machine, and the world never looked back.[1] Because oil was viewed forever after as being critical to success in wartime, nations began jockeying to control the world's known oil reserves. In the period around WW I, the oil-rich areas meant mostly Texas, with the Middle East known to have some potential as well.

The automobile concurrently sent the world into petroleum infatuation. In 1900, only a few cars existed, fewer than ten thousand, which were the playthings of the rich and eccentric. Some ran on electricity, some on steam, and some on derivatives of oil. By 1920, there were over 20 million cars and oil had scored a decisive victory as the fuel of choice, and the world's big shots began to see how critical it was becoming to ensure supplies of petroleum.

In the 1930s and '40s, some of the world's biggest oil fields found to date were discovered in the Middle East, and the great global carve-up began. The US, being a major economic force and growing at an unprecedented rate, was most aggressive in looking under every global rock to secure cheap supplies of oil. US oil companies, having pioneered the business in American oil fields, led the global charge.

After World War II, peace and prosperity arrived in most places, except the Soviet Union where only bags of onions and cabbages arrived (communism was hitting its dull grey stride), and the demand for oil skyrocketed. The world's free market economies boomed, even in (or perhaps partially because of) rebuilding countries like Germany and Japan. The American economy was roaring as well, and the US cemented its position as the undisputed financial powerhouse. Underpinning this whole frenzy of global activity was cheap and plentiful oil. The automobile phenomenon had spread worldwide, and travel/freight systems developed based on this cheap miracle fuel.

The search for oil accelerated, becoming both more frantic and more global. The US-led industrial juggernaut reached out across the world, and many an unsuspecting undeveloped country received a visit from the hulking international equivalent of a vacuum cleaner salesman, except the door-knocker in the suit wasn't selling, he was buying.

The primary door-knockers were the big oil companies at the forefront of the petroleum revolution. The Seven Sisters were largely American; the group came to life with the antitrust smashing of the US Standard Oil Corporation, which created some of the sisters, and the rest were born through the Consortium of Iran. I'll stop there; you can find better histories of the Seven Sisters elsewhere.

Finally, this is where it starts to get interesting with respect to big oil. The full-court-press global oil race helped put in place deals and structures that caused repercussions and upheavals so huge that they have created much of the instability in the world today. It took centuries, but something finally knocked religion and politics out of top spot on the global "reasons to go to war" list.

YOU GOT OIL? WANNA MAKE A DEAL?

Big oil's expansion in pursuit of global petroleum reserves was a sign of the times, and they pursued it as it was done in those days: they buddied up with foreign governments. It is important to remember the context of the times, with colonialism then being a far-from-dead concept.

The context of significance was that, at that time, most of the nations that turned out to have huge oil reserves tended to be the industrial equivalent of hillbillies. The door-knocking global corporations were like a cross between old-fashioned elixir salesmen, slick lawyers, and real estate developers all rolled into one formidably charismatic arm-twister.

As a predominant commercial development theme, the persuasive deal makers wanting to meet the head of state and cut a deal were well aware of the value of petroleum reserves, while the various local owners were often as ill-equipped to deal with this commercial onslaught as were Manhattan's natives.

The biggest problem was that when the oil companies came knocking, they weren't dealing with robust democracies that could act in the best interests of the people of the nation; there were strong crony loyalties to worry about first. Thus, the oil companies struck up deals with some often-dubious characters who happened to be in power that fine day.

The question that company representatives asked when they landed was, "Who is the government?" not necessarily, "Is the government horrible and corrupt?" because the latter question is a difficult opening gambit when trying to land oil exploration concessions. Fundamentally, the oil companies weren't particularly interested in the well-being of the people of these regions; those people were simply objects that had to be moved in order to get to the good stuff underneath. Stand back everyone, that there is a seismic crew, and behind them is what we call a drilling rig. What we need from you locals is a safe place to stay, and if you could find us some food that would be great. Western food please. That will be all, thanks.

Looking back now, it's easy to view this as colonial and arrogant and unfortunate, but at the time, making such judgements may have been a little murkier. Recall that only a generation or two before that, European countries had been busy colonializing every piece of land they could find. These days, it's hard to imagine what that was like, but a scant several hundred years ago, it was commonplace for a European country to sail up on some unsuspecting island, scout around for other Europeans, attack if some were found, and if not to simply pound a flag into the ground and declare it their territory. It seems surreal and ridiculous, but that's the way it was. For example, look at a chain of beautiful Caribbean islands that had their own cultures and heritage that are now neatly cleaved in two, with one set dubbed the British Virgin Islands and the other the US Virgin Islands. Thanks for stopping by. Or even more bizarrely, consider St. Maarten, a beautiful island smaller than most cities that is almost comically divided, with one side Dutch territory and the other French. It is the most beautiful and absurd place I've ever seen, with one side of the tiny island speaking French and serving croissants and driving Citroens, and three miles away over a few hills, it is Dutch to the core.

While it may appear an embarrassingly colonial and shameful example of western/European arrogance, it's hard to say if the whole phenomenon was good or bad. It is easy to see the bad side, that of a stable population overrun by invaders who simply took over. However, there may have been silver linings that offset the negative to some extent. Perhaps locals were starving, and the European connection brought supplies in exchange for letting them put their stupid flag up here and there. Perhaps the locals were bored and welcomed the novelties these odd foreigners brought, all the silly customs and clothes and pointless formality. Perhaps they were relieved to be protected from pirates, or perhaps they resented the hell out of the occupation. I don't know, I wasn't there.

The same may be true of the oil colonialists. Sometimes it turned out rather well, and sometimes quite badly. But put yourself in the shoes of those companies of the day. What was the right thing to do?

These companies were at the forefront of globally expanding the supply of oil for ever-more voracious western economies. Many of the countries that were found to have significant oil reserves didn't have anything resembling a democratic government, yet those governments were open to making deals. Were the oil companies to avoid a country because the government appeared to be nasty? Was that appearance correct? How stable was that government? Had it been good (or bad) for two years (or twenty)? Or six months? And how would you know from a few cursory visits?

Keep in mind the depth of knowledge of these foreign lands in the middle of the twentieth century. We tend to forget that knowledge of other countries and cultures was simplistic at best and consisted mostly of a few stereotypes. The resource material of the day seems laughable now. My mother was extremely proud of an investment she made when she was young; she bought a set of encyclopedias that occupied a sacred space on our living room's bookshelves. Growing up, this shelf of books gave us a leg up on other kids because they provided a window into the world that many of my classmates did not have. Even the school didn't have a set. It is almost embarrassing to think about that depth of knowledge now, and the singular view of its creation, and the frozen-in-time snapshot of what the author perceived as reality. I remember doing school research projects about other cultures and using those books as the sole point of reference. For any remote country, the precious encyclopedia had a handful of black and with photos which, as photos tend to, formed many of my underlying judgements and opinions about the place. Today, a five-year-old kid can street-view a thousand vantage points in that same exotic country, and through visual reference alone understand the place better than I ever did. The big oil firms may well have been as ignorant as I was; the local knowledge would hardly have been much greater than what I had from that encyclopedia set, and what was known would have been gathered through a filter of completely unknown quality. If the companies sent an intrepid explorer inland to find out what the natives were like, or what the cultural situation was, the entire analysis may have been completely distorted by the explorer's random encounters with exotic new alcohols or pretty women.

Big oil's pursuit of reserves therefore led them to these foreign lands perched atop sometimes staggering quantities of oil reserves that the locals had no interest in, and which the governments were more than happy to deal away. The oil companies worked with what knowledge they had and cut deals with whomever appeared to be in charge. It is possible that the big oil companies had some reservations about what they were doing, but would you turn your back on potentially massive oil reserves based on some hypothetical mini-drama that the local government promised to deal with anyway? Besides, one of the other competitors would surely have held their nose and plunged in anyway, so better to be on the safe side and cut a deal.

CAN'T GO BACK

Another factor to consider relates to the habits of the times. Remember that active colonialism had been completely acceptable not many decades before. The whole idea of landing in a foreign land with some tacit and thorough business/geological knowledge and cutting a deal with a bunch of unsophisticated bumpkins seems utterly preposterous in the standards of today, but in those days, that was how the world worked.

Colonialism, even in its waning years, left an indelible mark on much of the world. It is too overwhelming to attempt to evaluate whether that was good or bad. Is the island of St. Maarten better off or worse off because it has been divided into a Dutch side and a French side and irrevocably marked with European culture? On one hand, the very name of the place is Europeanized, meaning the original culture has been more or less wiped out except for the pesky humans who somehow remained through it all. Or are they better off because the European influence brought medicine, European educational influence, commerce, and cruise ships?

Would Saudi Arabia have been better off if it had developed its oil resources on its own? Or wait, that question glosses over a vital issue. Would Saudi Arabia have been able to develop the resources on its own? Control is nice, but capital can be pretty sweet too. As with colonialism, we'll never know if those developments led to more pluses or minuses, but the question of whether the petro-colonial developments were good or bad is at least a debatable point.

It is therefore hardly sensible to apply today's standards of propriety to what was happening in those days. We can't go backwards and imagine what it was like to be in that race for petroleum, to be dealing with countries that had little or no knowledge of industrialization, and that were open to dealing away what would become national treasures. There is no point in punishing big oil for what happened then. Was it terrible? In hindsight, probably in some instances, but we also can't say for sure how those oil-rich nations would have fared had they eventually developed the oil on their own. Perhaps all the wealth would have wound up in the hands of a few, because democracy didn't seem to be a thing then, and there may have been eternal wars and power struggles and squandered resources. Maybe developing countries would have had a dozen dictators with a billion dollars in a Swiss bank account instead of only two. It is hard to say.

THE MARK ON PEOPLE

The consequences of this are worthy of our attention in the current environment. It may have been horrible that one of the Seven Sisters bought off a corrupt government in some foreign third world country, and some families were unjustly deprived of their ancestral homes in the pursuit of petroleum, but some aspects of society we can't redress. Would the natives like to buy back Manhattan for, according to Wikipedia, $1,050, which is apparently the value today of those sixty guilders or whatever was paid for the land? I suspect they would. The issue of course is, where does the attempt to redress things stop?

That is a huge issue for another day and another book, or a hundred books, but for this discussion, it is imperative that we understand some of these factors at work today. The oil patch can sit back dumbfounded as to why it is vilified in some circles when it provides such a life blood to society, and that view is valid through one particular lens. The oil these companies produced went into the spiral of economic development that led to the gloriously cushy lives almost all of us lead today.

But maybe one of those protesters holding a sign is a direct descendant of one of the families from some small, developing country who was kicked off their land by a multinational oil company. Maybe since then the person has led a nomadic existence as a refugee, or worse, and now finds themselves in North America or Europe, with the chance to raise their voice against the biggest evil they ever encountered in their life, and they're doing so with a vengeance. Maybe the guy on the train that I see some mornings, who appears to be of African descent and has three parallel and brutal scars on each cheek, from ear to mouth, has a story to tell about how he came to Canada that would make my hair stand on end. Maybe the roots of that would be some resource-driven conflict I could never understand unless I lived through it.

Who's right in that debate?

It doesn't matter who's right. What matters is this: Around the globe are plenty of people who have been either personally screwed by such deals, or are descendants of people who were. Some elements of history may be subject to subjective revisionism, but this one is definitely not: if your ancestors had, through no fault other than naiveté or innocence, been party to a trade so lopsided that it became legend, whomever tries to get you to sign the next deal is going to be entering the ring with a wildcat.

This is exactly the state of most North American native relations and land claims. There is no need, oh lord there is no need, to go through the whole sordid chain of events. It would require far too much energy for me to even think about. As with other aspects, what we need to remember is that those natives are of the mindset that their ancestors were taken to the cleaners, and there is no way in hell they'll repeat that mistake. On the other hand, these same people have the task of defining what position they want in the present world, and how to work within our global energy situation, because the old ways are not coming back. It is not a simple issue.

THE MARK ON GEOPOLITICS

At the beginning of the 1970s, as was noted earlier, the Seven Sisters controlled 85 percent of global oil reserves. By about the turn of the century, it was 6 percent. There is a reason for that.

Back in the 1950s and '60s, the large international oil companies buddied up with a huge spectrum of nations to take the eggs from the nest, or more accurately, get the nations to sign papers allowing them to take the eggs from the nest. This makes it all sound nasty, but to be magnanimous, it is possible that these multinational corporations made a lasting contribution to the countries other than armed compounds where ex-pats could feel safe and drink familiar beer. The development of these resources offered many nations a chance for economic development that may not have arisen in any other way.

History would indicate that most of these nations weren't thrilled though, which is one of, if not the biggest, reasons OPEC was formed. The Organization of Petroleum Exporting Countries was created when a number of developing countries that were responsible for a lot of the world's production decided they wanted to control their own destiny. Multinationals were kicked out and state oil firms took control of petroleum reserves.

The US and the Seven Sisters had no choice but to play along; however, the playing field of the world has been forever tilted because of oil. We can see that to this day, based on whom the US is friendly with, and why.

All the machinations of global geopolitics are too complex to piece together here, and possibly not worth trying to at all. I'm in the camp that believes that we only know parts of the stories and not much at all about the true motivations and drivers of events as they unfold. Only the people sitting in the decision-making chairs know all the factors at play; we simply observe the outcome and judge it through our own filter. We tend to think that something huge and material like, say, an international trade agreement would be based on mountains of information, analysis, and dripping with cleverness. On the other hand, those assumptions may be wildly off the mark, because events actually unfolded along planes that may seem nonsensical to bystanders but are tactically preferential to those in power. Maybe two negotiators liked each other and went golfing and signed a deal based on that. We'll never know all the details.

As a case in point, on September 11, 2001, twenty terrorists seized control of four airplanes and attacked the continental US. Like you need reminding. This was, to put it mildly, a very big deal. The US is not likely to opt for pacifism in these situations, rightly so, and no one was surprised when a few days after the attack the war drums began beating.

What unfolded in the aftermath though was an example of geopolitical machinations that appeared then, and do to this day, to be insane, but they actually happened. Of the twenty lunatics that attacked the US, fifteen were from Saudi Arabia and the rest were from the United Arab Emirates, Lebanon, and Egypt. The US snapped into retaliatory action and invaded Afghanistan, and then Iraq. In Iraq, the US sought to topple a dictator that had been an ally of the US several decades before. In Afghanistan, the US battled some extremely hardy mountain rats that had received some impeccable survival/military training from none other than the US, who had trained the Afghans to defend against invading Soviets some years before.

On the surface, these invasions seem nuts, because no attempt was made to go after the originators. But perhaps there are stories behind the scenes that we'll never know. History and military buffs endlessly analyze these things, and probably one of them out there gets it exactly right, simply through the law of averages. We may never know which one is correct because many theories are plausible. What we can surmise though is that resources, in particular petroleum, almost certainly were a big factor in deciding whom the US went to war with.

Consider Saudi Arabia. Not only did some of its citizens bomb the US, but the country has been until recently one of the most regressive dictatorships on earth. Women weren't allowed to drive or even jog in public until recently, and the list of human rights abuses ranks the country right up there with the world's barbarous best. The US has toppled more than a few governments for far fewer reasons, or helped to anyway, yet Saudi Arabia is always welcomed as a friend. You can guess why; the existence of a claimed 260 billion barrels of oil under its Gucci loafers has more than a little to do with it.

But it goes deeper than that. Iraq had a lot of oil too, but it got the hammer. It got "shocked and awed," as the US military put it. The alleged reason for the invasion—that Iraq had weapons of mass destruction—turned out to be untrue, and actually seemed ludicrous from the start. So, there were probably other factors involved, and oil is likely one of the major ones. Consider that Iraq has a lot of oil and was attacked, and that Pakistan has weapons of mass destruction but little oil and was left alone, even after Osama bin Laden was discovered hiding there.

We therefore have these cataclysmic events that happen in global geopolitics that seem inexplicable, and quite often oil is at the heart. When geopolitical events happen and oil is involved, the world's supermajors are usually not far away. This isn't necessarily a critique of their ethics, because to some extent, only the supermajors had the expertise, capital, and infrastructure required to develop those resources.

Those tools of big oil then often landed them right in the middle of global geopolitics, with a fascinating tension arising. As a supermajor on the hunt for mega-discoveries, in a competitive search that was spanning the globe, which master did one bow to? Would it be capital markets, who handsomely rewarded the profits these discoveries could yield; the US government, eager to secure if not ownership at least a commercial connection to oil reserves in otherwise dubious countries; or the governments of those often-nefarious countries themselves? In other words, if a deal could be cut with a foreign despot in order to access oil in their country, should the supermajor turn a blind eye to the frequently unfriendly manner in which these despots cleared a path for the supermajor? If a tribe was kicked off its land so that oil reserves could be developed, should big oil walk away because that wasn't a nice thing to do?

Many of those "friendly-access" side deals and agreements came back to haunt big oil, because times have changed. Fifty years ago, when a tribe was kicked off its land so that a new oil development could happen, big oil was confident that that would be the last they'd see of that colorful bunch of natives. Ironically, thanks to globalization and cheap energy, which has permitted affordable international travel and migration… surprise, look who's here!

THEY WALK AMONG US. ACTUALLY, THEY ARE US

If your parents or grandparents are typical citizens of the western world, ask them how many Vietnamese they knew growing up. Or Russians. Or Ethiopians. Or Japanese. You get the idea.

Now ask your children the same thing. You will most likely get a list that reads like a United Nations committee meeting roster. Global travel has cracked the world wide open. We now pass people on the streets who grew up in the Congo, followed by maybe a Russian or Iranian or Argentinian, and racists find their job getting harder and harder, since exposure to other cultures tends to undermine the certainty of ignorance.

This intermingling phenomenon has brought many multicultural benefits, a lot of delicious food, and brought to our world viewpoints that we never knew existed before. The idiosyncrasies that formed real life for a person growing up in Nigeria would have been unknown to us fifty years ago, and our perceptions of such foreign countries would have been formed by small snippets of pictures, information, and ignorant stereotypes. Now, we may easily encounter someone who has lived through all sorts of horrors, and they can recount them to us first hand.

We can't ignore the influence of big oil in the views and beliefs these immigrants bring to our shores as well. We may have had a jovial idea of big oil growing up, how ExxonMobil/Imperial Oil had that super-cute tiger as a mascot. Who would grow up hating that? Doesn't it want make you want to stop there for gas, Daddy?

Maybe a person from Africa would not be so interested in dealing with the big friendly tiger if that same company had cut a deal with a dictator to explore for oil in your ancestral territory, and that dictator dealt with protests in a brutally efficient manner before they even became protests. Maybe that young man or woman escaped as an orphan or if lucky with a parent to North America or Europe and brought, perhaps, a wee bit of mental baggage with them.

When people these days have a hatred for big oil and decide to join the fight to end the dominance of fossil fuels, keep in mind the path they may have taken. Maybe they or their family suffered from injustice handed out by a ruling despot who'd cut a deal with an oil company. Maybe one of their ancestors passed down an old blanket as a reminder of some certain key event or bad trade that happened a few nodes back in their family tree.

Some protesters are of course less than admirable, but we should think twice before thinking they all are.

On the other hand, it can't be ignored that the cheap energy that the multinational oil companies provided has permitted the world's citizens to travel so freely. Without cheap energy for travelling, perhaps many more cultures would remain isolated from each other, and we would never have learned the wonders of sushi or pad thai.

Now, wrap this all into the modern drive to get rid of fossil fuels. We can see where the waters become muddied with respect to the debate of how much we need fossil fuels versus how much we need to get away from them.

8.
What about coal, the "other" fossil fuel?

Imagine you're a dentist. No one likes you, but you take every Friday off and you have five Ferraris, so there are perks. You spend your day rummaging around people's mouths, deftly handling the simultaneously delicate tasks of being both nice to the client while at the same time pointing out that you could make them far more attractive if given half a shot.

One day, after a grueling five-hour shift, you're sitting in the east wing of your third house, the handy one close to work, philosophically pondering life while enjoying a glass of your favourite beverage that costs as much as a Hyundai. You think about how unfair it is, how even children spit on you as they walk by, yet modern life would be difficult indeed without the services your profession provides. Dentistry is not a necessity of modern life, but it is close. It is a hallmark of the advancement of civilization, where a lifetime of aching jaws, abscesses, and old-age single-tooth-grins are wiped away by a few unpleasant but brief hours in a reclining chair and a periodic good old-fashioned wallet cleansing. Is that not good for society as a whole? No matter how black-hearted and cruel your profession may be, do you not perform a function that is a necessity for modern life in many parts of the world? Is it too much to ask for a little respect?*

---

* SIDE NOTE ON DENTISTRY FIXATION: My pathological and perhaps slightly exaggerated view of dentists stems from a brutal experience as a young child that I only remember in snippets, after which experience the dentist apologized to my mom for not freezing me properly. My subsequent memories of dentists therefore bring either abject terror or, now that I'm a bit older, a fair number of admittedly unpleasant and fortunately wildly impractical "procedures" I'd like to perform on them. Sharp-eyed readers will perhaps notice this slight unconsciously-injected bias against the profession, which I apologize for but am helpless to stop. My sincere hope is that I am never forced to, say, have to face a

THE END OF FOSSIL FUEL INSANITY

Welcome to the world of coal.

IT'S DISGUSTING, BUT EVERYONE'S DOING IT

The movement against fossil fuels has one unfortunate aspect: it tends to lump all the fossil fuel options together as a single entity. Consider that for a second. At one extreme of the family is an invisible gas—methane or natural gas—that burns with a blue flame and is clean enough to be considered an environmentally friendly fuel (which is also produced naturally in swamps and things that have been at the back of your refrigerator for eighteen months). The other is synonymous with not getting any Christmas presents: black, dirty rocks that are best known for making coal-miners look like negative images of raccoons when they come out of the mine and that emit foul clouds of black soot when burned. One is gassy and light and clean burning and the other is dark and mean and Dickensian; like the difference between a happy kindergarten teacher and a dentist.

In one major way, that's not fair at all, just as dentists are unfairly treated by some of the lesser but more dogged members of society. As with many other things discussed in this book, coal is undeniably responsible in large part for the sumptuous lifestyle we enjoy today, where minimum-wage workers have access to a wider array of luxuries than royalty did five hundred years ago. Never before in the history of humankind has it been so easy to sustain oneself with so little effort as today and, disgusting or not, coal got us here.

That conclusion may seem outdated and pertinent mainly to the nineteenth century, and it is, sort of. That was the heyday of coal, the era when it was universal and dominant. But another critical element of today's fossil fuel wars is that people forget just how relevant it is even to this day. Everyone is indeed doing it, except for a few jurisdictions at the cutting edge of the green energy movement, and they have only kicked the habit because they have other plentiful options (as in other fossil fuels, or bags of money derived from fossil fuels—hello, Norway).

---

moral dilemma such as encountering a dentist in a time of extreme need, such as a car accident, where I fear the full wrath of several decades of pent-up scheming and raging will rain down in one ugly and spectator-perplexing minute. Apologies for the diversion; now back to coal.

WHAT ABOUT COAL, THE "OTHER" FOSSIL FUEL?

THE ROLE OF COAL

Before the harnessing of electricity on a large scale, sourcing heat or power meant burning stuff. It doesn't take much imagination to see that burning wood for heat and cooking can only support a small population; any decent-sized city that tried to get by burning only wood would devour surrounding forests like a plague of locusts, and the wood smoke on a winter's day would be thick enough to eat. The primary advancement of civilization out of the middle ages then was the result of having something to combust other than wood, a breakthrough that afforded a whole lot of other interesting things to happen also.

The industrial revolution was a confluence of advancements in manufacturing, transportation, and communications, all made possible by having relatively cheap energy, which basically meant coal (until petroleum came along). If you're interested, the industrial revolution has engendered about 14 billion books; full details are better obtained elsewhere.

The point of significance to be made here, rather than talking about, say, the first steam-powered toaster of 1862 or some other comparable point-of-excitement for historians, is that coal has been the go-to development fuel for much of the world that achieved or is striving towards western levels of industrialization. Today, when we talk about growing industrialization, only two countries really matter: China and India. Sorry if that sounds demeaning, Rest of the World, but it's true. No other country can hold a candle to the energy requirements of these two, with the way they're growing. The two nations have a combined population of 2.7 billion people, which until recently were pretty much at the same stage of industrialization as the west was a hundred years ago. But they have been catching up with a vengeance. China has become a global economic powerhouse, and a voracious consumer of, well, everything. India is behind, but on the same trajectory.

They have done this, and continue to do this, on the back of coal. Yes, there are grand plans in both countries to move to renewables, and there is frantic activity to do just that. But as with many other things in the energy world, we hear the statistics about new solar installations, which are clean and new and awesome, but we forget that, as recently as 2016,

India got eighty percent of its power from coal, as did China. That is what development looks like.

As the economies of these two nations advance, they will continue to make their energy portfolios greener, but consider what a herculean task it will be to get those coal-usage percentages down to even half of present levels. Remember also that these are two of the fastest growing economies in the world; getting off the coal addiction means not just converting all coal usage to renewables but not relying on coal at all for any growth from here forward. In other words, the very idea is not possible in the next few decades. You will hear a lot about Chinese (and probably soon Indian) cities banning internal combustion vehicles at some point. That is not the final leg of the green journey but merely the first, and is being driven by sheer survival. Do a web search for images of Chinese smog and you'll get a murky view of what's driving that transition.

HOW COME COAL GETS A FREE RIDE WHEN IT'S SO DISGUSTING?

There have been more than a few headlines recently about how China is going green. The country gets kudos for solar installations, for promoting electric vehicles, for speaking of a plan to ban internal combustion engines within several decades. And the crowd goes wild.

Meanwhile, a chart of the actual quantity of emissions that China is pumping out would make the board of ExxonMobil feel like Greenpeace. If global warming is truly a human-made phenomenon caused largely by fossil fuel emissions, China is, through nothing more than the staggering impact of 1.4 billion people striving for middle class, pumping out enough greenhouse gases to single handedly boil the planet.

Yet paradoxically, China has been gathering rave reviews for its environmental performance. These high-fives have been because coal's share of the energy mix decreased over the past number of years. Yay, shouted the scorekeepers. See, if China can do it, why can't everyone? All you fossil fuel fools out there, look how they did it by adding electric vehicles, solar installations, and wind power. Wind power in particular makes headlines because of the sheer number of turbine installations added. Wikipedia points out, "In 2016, China added 19.3 GW [gigawatts] of wind power

WHAT ABOUT COAL, THE "OTHER" FOSSIL FUEL?

generation capacity ... and generated 241 TWh [terawatt hours] of electricity, representing 4 percent of total national electrical consumption." And China has "pledged to produce 15 percent of all electricity from renewable resources by that year [2020]."

Sounds incredible, right? Now *that's* how you save the planet.

Oh wait, there's a few more paragraphs here. Yes, China did drop coal's share of the energy mix year over year—*from 64 percent to 62 percent*. That still includes some 3 billion metric tonnes of coal per year, and power consumption is growing by almost 4 percent a year. China is building more coal fired power plants, planning to add 140 GW's worth of coal fired capacity, which as you'll note is seven times as much wind power as was added last year.

In other words, good on them for the green energy work, but the whole media frenzy about this progress is an absolute sham. According to a NY Times article,[1] "Even if American emissions were to suddenly disappear tomorrow, world emissions would be back at the same level within four years as a result of China's growth alone."

The amount of coal being burned is responsible for a quarter of greenhouse gas emissions, but you see no one protesting coal usage. That absence is incongruous with the attention paid to other perceived climate villains. Perhaps you've heard of the evils of the Canadian oil sands. Everyone has heard of this petroleum deposit, from the monks of Tibet to the ring-tailed lemurs of Madagascar. Not one of these creatures has a favourable opinion; without exception, they would know for a fact that the oil sands are destroying the planet.

However, according to Statistics Canada, the oil sands produce about five percent of Canada's greenhouse gas emissions, and Canada produces two percent of the world's total. In other words, the oil sands produce 0.1 percent of global greenhouse gas emissions while producing 2.5 percent of global oil production. China's coal plants currently under construction will by themselves dwarf greenhouse gas output from the oil sands.

The lengths by which China has gone green are not much beyond sheer survival. The World Bank estimates air pollution kills over a million people per year in China, according to a study done by The Lancet and reported in the NY Times.[2] The push for electric vehicles by the Chinese government is therefore not much of a "save the planet" thing, it is a "save grandpa" thing.

I'm sure China would love to go all green, to have nothing but EVs and to stop burning coal, but yet again, we see the intractability of changing the energy system. We can see everywhere we look how hard it is to change it in any meaningful timeframe, and how hard it is to integrate new growth into existing systems.

China's necessity to promote EVs and cut back on any major smog contributors is a near-mandatory response to changes in population, changes in growth, and changes in the way the country must heat, feed, and move these growing populations. The green push may make headlines and thrill hard-core environmentalists, but it is nothing more and nothing less than keeping China's largest cities livable. The nation is doing what is necessary for its population's survival, and EVs are one part of the solution, just as replacing coal with natural gas is. The rest of China continues to burn whatever it can get its hands on, and its emissions are the world's largest and the biggest contributor to global warming by far. Coal remains the most dominant fuel in China, and will be for some time. The same can be said of India, though that country is less advanced on the renewable energy front so gets fewer headlines.

So, in the end, what about coal? Does it get a free ride? Well, it manages to get by without the massive and persistent protests that petroleum and petroleum infrastructure does, so in some ways that's a free ride. But at the end of the day, coal is simply too important to the world at the present time to do away with.

More importantly, just like oil is produced in some places and consumed in others, fossil fuels like coal are used in one place but the protesters are in others. That is, North America and Europe have many protests, but the world's main CO_2 problem, the burning of coal, is happening relatively peacefully on the other side of the world. Protesters are against what they see right in front of them, the elephant and the blind person's problem all over again. It makes sense to fight a Canadian pipeline, because they can and it's right there in the heart of a major city. It makes no sense to protest Chinese coal consumption, because it will do no good whatsoever.

WHAT ABOUT COAL, THE "OTHER" FOSSIL FUEL?

This is another facet of human nature that we simply must live with: our world is simply what we see. We don't protest coal usage on the other side of the planet, even though that would be a hundred times more environmentally beneficial, because, ah, that looks like a total pain in the ass to even try.

With these discordant protests, where people—in the name of climate change—try to halt things that are far less harmful to the climate than others they ignore, we see another brick added to the wall between those who provide energy, and those who don't.

9.
Rise of climate change as a global issue

Earlier, we touched on the topic of climate change and how our understanding of the climate improved at roughly the same time that we realized the average temperature was rising. Some of you may not like the sound of that, those that challenge whether the climate is changing and why, but for purposes of this discussion, we're going to go with that widely held viewpoint. The purpose of this book is to hack through the jungle of obstinacy to find a path forward. As a conciliatory starting point, let us assume as true the notion that the globe is warming and that it may very well be related to human activity.

When the conclusion that human activity is responsible became widely accepted, the seeds of the climate change movement were germinated and grew into mighty trees. Apparently, they were amply fertilized too—no comment on what makes good fertilizer—because the climate change movement is now the largest on earth.

FORTY YEARS AGO, THERE WAS TALK ABOUT ANOTHER ICE AGE, RIGHT?

The very mention of the 1970s ice-age-is-coming story will make apoplectic the average climatologist, and he/she will throw this book into a river. To even bring up the discussions from that era drives the modern climate industry crazy.

Regardless, the story needs to be mentioned because it was part of the narrative, even if only for a short time. No one back then laughed or ridiculed the notion; it was simply another in a long line of "look out the sky is falling" news stories. The piece is, however, now laughed and ridiculed mercilessly, which is also a part of the story.

Nowadays, it sounds bizarre to stand up and say that, according to the tea leaves, the world was possibly headed for a dangerous cold spell. But that story did indeed happen in the mainstream media, with a particular 1975 Newsweek article[1] at the epicenter of it. At the time, it was a reasonable interpretation of what the world knew about the climate and how it was changing. The interpretation of the data wasn't particularly contentious either, because the current earth-is-warming-dangerously movement was not a movement at that time.

Today however, to even acknowledge the article is to poke a sharp stick in the eye of the current climate business. They don't like it one bit. A 2014 Scientific American article[2] had this to say about the 1975 story: "Temperatures have plunged to record lows on the East Coast, and once again Peter Gwynne is being heralded as a journalist ahead of his time. By some... Gwynne was the science editor of Newsweek thirty-nine years ago when he pulled together some interviews from scientists and wrote a nine-paragraph story about how the planet was getting cooler... Ever since, Gwynne's "global cooling" story—and a similar Time Magazine piece—have been brandished gleefully by those who say it shows global warming is not happening, or at least that scientists—and often journalists—don't know what they are talking about."

It is worth putting this quotation into context. This is from *Scientific American*, of all places, a former bastion of apolitical, objective, free-range science thinking. Note the menacing and completely unscientific undercurrent, the story's sub-header that states, "Nine paragraphs written for Newsweek in 1975 continue to trump forty years of climate science," and how Gwynne's piece has been "brandished gleefully by those who say it shows global warming is not happening." There are of course also much more critical pieces to be found online, ones that mock the story as being hysterical and ignorant, and go to great lengths to discredit it. The author himself was coaxed out of retirement to hurl feces at those who actually gave credence to his story.[3]

In Gwynne's more recent article where he chastises those who use his original as a weapon, he makes some good points as to why the original may have been premature, but he also recreates the mistake that he says invalidates his earlier comments. The original, he points out, was based on observations at the time which clearly were not well understood. Now, he asserts, they are, and he quotes the National Academy of Sciences to prove his point. It is somewhat problematic to the success of this backpedaling that in the original piece he also quoted the National Academy of Sciences, which was then not at all afraid of publicly musing about the real possibility of global cooling. But, you may ask, what's so bad about that?

That repeated quoting of the same source with two entirely different outlooks brings up a crucial point: The National Academy of Sciences viewed global cooling as possibly legitimate *at that time*. As was pointed out earlier, the science continues to evolve and not everything is settled and carved in stone just yet. The analytical tools available now are light years beyond what was available forty years ago, and the tools of 1975 were toweringly capable compared to those of forty years prior to that. The "science" was evolving in 1975, and it is now. Why is Mr. Gwynne convinced that now the knowledge base is final and complete? Would he not have thought the same thing in 1975?

One of the reasons this continual metamorphosis is problematic was well documented by Gwynne in his follow up piece when he quotes NASA bigwig Gavin Schmidt: "'There's too much hand waving in science journalism,' Schmidt noted. 'Scientists don't spend a lot of time when talking to journalists about what their research doesn't mean. One of the fault lines between science and journalism is how you pull together the bigger picture. So a reticence on the part of scientists to fill in the big picture, and over-enthusiasm on the part of journalists to say what does it all mean, means that the journalists don't get it quite right.'"

It isn't absurd that the global-cooling story actually had credibility. At that time, climatologists did observe a multi-decade cooling trend. The major environmental concern was depletion of the ozone layer, and in the exact same vein as we see reported today about climate change, stories about the impending perils to humankind if ozone continued to be depleted were rampant.

In the 1970s and 1980s, a substance known as chlorofluorocarbons (CFCs), widely used in aerosol spray cans, was found to be damaging the earth's ozone layer. The ozone layer traps harmful ultraviolet radiation, but it also has the effect of trapping heat near the earth's surface and keeping it toasty. If the ozone layer were depleted, there was (at the time) legitimate fear that the earth would cool.

In the 1980s, a British expedition to Antarctica stunned the world with the announcement that they'd discovered a giant hole in the ozone layer[4] and that the risk to humanity was severe and growing rapidly unless we dealt with the problem immediately (sound familiar?). The countries of the world made remarkable progress in tackling the issue, culminating with the Montreal Protocol of 1987, a universally adopted agreement to wipe out use of CFCs (familiar again, right down to the French connection). It was highly successful over time, and now the ozone layer has begun to heal.

I am not equating the risk from loss of ozone with the risk of climate change. The point is that the global cooling story was not ludicrous, *at that time.*

There are valuable lessons in stepping back and watching these narratives unfold. Very different stories emerge when snippets are used to advance agendas, no matter by whom. This is of course a universal issue, not restricted to climate change. We need to be cognizant of the latest scientific developments and pay heed, but we shouldn't bet the farm that they are the final word in the story. If we were actually listening directly to the scientists we might never get to these crazy stalemates, but with every scientific observation being reported through the fun-house mirror of modern "what-does-it-all-mean" journalism, the stories become whatever we want them to become.

WE'RE CRAPPING ON GLOBAL WARMING THEORY THEN, IT SOUNDS LIKE

Nope, not at all. The globe may indeed be warming due to human activity. Hold on, let's look at this from a different angle by hopping over to the world of astrophysics for a second.

Astrophysics is an astonishing field, peering off into the great unknown where clues to the universe and our existence reveal themselves tantalizingly slowly like some gargantuan nerd-dream strip tease. It is an esoteric field full of beauty and wonders and keys to the existence of life as we know it.

Before getting too lost in the wonders of the cosmos, hold that thought. We need one small sub-analogy. Imagine growing up in a prison cell with no windows, where food was mysteriously dropped in every day. It's the only world you know. After a number of years, you would know your "world" quite well. You'd have inspected every little nook and cranny. Then one day, suppose that you are mysteriously freed, and someone places you in a forest. As you enjoy the immensity of your new space, it would dawn on you that you knew nothing of the world, that you now see that it is infinitely more vast and interesting than you'd ever imagined when all you knew was the inside of your cell. You would discover the wonders of seasons, clothes, and oceans. You even see what appear to be footprints and what appear to be signs of other life, which is mind blowing in its own right. Each new discovery would involve a mental reset and reframing of existence, and at each upgrade, you'd go from being a master of your little world to utter bewilderment.

This, in a nutshell, is astronomy, except the physical relocations are upgrades in telescopes or higher mountains upon which to build them. Imagine that you're God watching those crazy little humans doing everything they do, shaking your head and wondering just why in the name of all that's holy would they ever do *that*. But anyway, you're watching the humans observe the galaxy from their singular viewpoint, and the puny humans are thinking that they understand the whole thing. They find a new star and rewrite their theory of how everything came to be. As God sitting up there, you marvel at the simplistic naiveté that thinks the universe is understandable from a single viewpoint on a single planet. Not just a singular viewpoint but a tiny narrow one, because the vantage point of the universe from a single location is as absurdly narrow as standing in the middle of a wheat field in the North American great plains and thinking you can understand the whole continent from what you see. Then the humans upgrade their tools. They build a new telescope on top of a mountain in Hawaii, and they're excited because the new view invalidates many of the prior intergalactic theories. Now the lunatics think they have cracked this cosmic nut for sure because they can see things *so* much better, and as God you're looking down at this little fly-speck of a planet, eating some really good God-snack and thinking about pointing out to them that even with their fancy new telescope their vantage point changed—from an interplanetary perspective—by such a tiny amount that even He can't measure it, but you've been down this path before of trying to point out the obvious to those boneheads so you give up and just go see what's on God-TV.

Before you get too comfortable in that amazing mental God-chair, go back for a second to the cell dweller. At what point of his/her outward journey would he/she be comfortable saying okay, I finally get it, I understand the world?

At what point does an astrophysicist say okay, I get it, I finally understand the universe?

At what point does a climatologist finally say okay, I get it, I finally understand the climate?

With that said, just because the climate and the interaction of humankind may not be thoroughly understood, it does not mean that we should by default argue against climate change.

CLIMATE CHANGE DENIERS

The tag of climate change denier is, to be blunt, a truly abominable abuse and co-opting of the catastrophic misery of others. To utilize and ride on the back of the horror that was the Holocaust is a pretty despicable thing to do, and make no bones about it: the derogatory name is meant to create a connotation between those fools who deny that the Holocaust took place and those who think the climate is not warming. The usage of this term is, sadly, a commentary on how the debate has sunk into the muck.

Having said that, some people are greatly irritated by some of the parameters the climate change movement represents, and thus vociferously attack climate change itself. This unfortunate side effect of the fossil fuel wars goes a long way to explaining the ruts we're in, not just the thorough lack of progress but also in the obstinacy of the opponents.

The debate has been tainted with the same poison as politics. In that disgusting realm, participating usually means taking sides, even if by default. What that means is that siding with one party means being caught up in the whole thing, and as a member of that party, you are painted with the brush of the lowest common denominator of your party's membership. If one of your fellow party members craps on an opponent with some vitriolic nonsense, then you did too, because you march under the same flag. And then the downward spiral begins, because you will get crapped on by the opponents, who see you as an interchangeable member of Party X, and you'll be like hey, I'm not that bad and they'll be like uh yeah, you are, and then it's all over. Civil discourse ceases, and any hope of progress stops.

Climate issues are now the same thing. The petroleum industry for a long time had its way with the world because governments wanted it that way and because the growth of fossil fuel usage seemed unstoppable, but with the rise of concerns about a warming planet, the industry went on the defensive. Some voices against the climate change movement were reasonable, but some were not. Then, just as with politics, the petroleum industry became synonymous with the lowest common denominator, meaning the worst behaving, least respectful entity. The environmental movement was happy fighting at that low level of civility, and the more boisterous and clever came up with clever insults like the denier tagline, which discredited the whole environmental movement in the eyes of many, and rendered the climate change conversation political.

SHOW SOME RESPECT

Some readers may wish for a definitive statement one way or the other, to hear about why climate change disbelievers are disgusting pigs or why the environmental movement is a hypocritical bunch of media manipulators. They won't get it here. There is too much work to do.

It's time for some grown up behavior. It's time for both sides to admit that we can and must reduce our usage of fossil fuels in a significant way in the not too distant future. It's time to admit that we currently cannot have our standard of living without fossil fuels, and that to get off them will not be easy. The task of transitioning to green energy is monumental; there is no time and energy to waste debating things like whether the earth was warmer eight hundred years ago. It is irrelevant.

To choose to fight against current observations and commonplace beliefs is a waste of time and energy, when we need to get to the same place anyway. We need to focus our energy on reducing fossil fuel consumption because it is not sustainable for a growing global population to count on it to the extent that we do. Consumption is the issue; it is important to not lose sight of that. Currently a great deal of effort goes into reducing reliance on fossil fuels by choking supply. This never works, and there needs to be a refocus there as well.

Some may be too far gone politically, on either end of this fossil fuel spectrum, to be interested in a solution-based path forward. This book is probably not for those people, although I hope they only realize that after they've purchased it. I am an absolute and total refund denier.

10.
Modern protesting: amateur hour no longer

PROTESTS THROUGH THE AGES—FROM THE NECESSARY TO THE NORMATIVE

Imagine you're a typical working-class person in the middle ages. Your life is going to be brutally short, and you're not exactly middle class—there is no middle class to speak of. Almost everyone is working-class, an appellation indicating you would be no more than one or two steps ahead of death. An over-cluttered garage is not on your list of problems. There are kings and queens and other forms of rulers, and below there is the class that is dedicated to either boot-licking to keep their heads on their shoulders, or scheming to remove the ruler's head from theirs. This class of semi-nobility and semi-powerful is the closest thing to middle class, but the gulf between their world and yours is gargantuan.

Regardless, in our example at the bottom of this heap is you, the commoner. Life is not particularly spectacular. A good day would see food on the table for you and the family, and a bad day could be as simple as getting a cold, because a cold could mean not working, and not working meant not eating. The risk of catching pneumonia is everywhere, which would probably be life threatening, and would leave your family destitute. The really bad days would involve stupefying acts of horror and bloodletting that we can hardly imagine but that were far from unusual in those days.

It would go without saying that in those days there were no protests. Life may not have been pleasant but it was simple—keep your mouth shut or become fertilizer.

THE END OF FOSSIL FUEL INSANITY

While some historian could probably unearth an example of what appeared to be a protest in fifth century Mesopotamia or something, and there were various European revolutions, those are irrelevant and beside the point because this isn't a historical deep-dive. The historian can have that victory because he/she obviously has so few. What we are concerned with here is when protests first turned into the modern variety, and for what reasons.

Perhaps the mother of all protests is the one that led to the formation of the United States (Wikipedia notes others in Europe in the few hundred years before this one; see paragraph above). We can also see how this worked as a turning point, that is, the mechanics of what made it possible. Had the US revolutionaries tried their shtick in, say, some part of England, they might well have been crushed quite quickly. Recall that there was no internet in those days. There were hardly even vegetables. An uprising within shouting distance of the ruler of the empire would quickly have been sniffed out and dealt with in an axe-y kind of way, whereas the Americans were distant and had a lot of resources, time, and rehearsal space that allowed them to hoist the middle finger to England.

The US was born of protest, and it quickly caught on as the in thing to do. The rise of global trade and the first stirrings of a middle class and associated freedoms meant that people didn't have to take as much guff as they had before.

Suppose you are now in the nineteenth century, and you're an actual middle class working stiff, because perhaps you have an actual job and are not out in the woods in a tug of war with a bear over a leg of venison. Things are a lot better than for your ancestors, but with not much for democracy on the horizon, the era of iffy rulers who claim power by force is still the norm. But now, thanks possibly to the US's success or simply because it was the cool thing to do, the general population was in some circumstances able to rise up against world-class tyrants. Not always, but sometimes, and the fact that this was now possible was a remarkable development in our history. It didn't always work out; communists took over Russia in 1917 in a rebellion against a dictatorship, and we know how that ended. But in the absence of democracy, the ability of protests to effect change had to be considered a good thing.

As democracy took hold in the west and morphed into the weird beast we know today, the protest movement shifted along with it. While democracy allows the individual to cast a vote and help shape a government, there has been for decades the sneaking suspicion that those politicians were and are answering to a far more subtle and powerful master than those yappy little voters. Thus begins the next chapter in the evolution of protests, where the goal is not necessarily to fight for survival, the right to vote, or basic human rights (although those protests are still happening and needed in various parts of the world), but to shape social policy for whatever cause one believes is doing the most damage and that simple elections don't seem to be doing much about.

Welcome to the sixties.

THE SIXTIES OPENED THE MIND AND SHARPENED THE PROTESTING TOOLS

The decades just prior to the sixties had little in the way of large scale protests as we know them today (the 1956 Hungarian uprising against the Soviet Union was a protest of sorts, but all it achieved was the opportunity for participants to look from close range down the barrel of a Soviet tank, so we'll put that one aside). The world was preoccupied with other things, like World War II, which was a substantial distraction indeed. There were of course protests against working conditions that formed labor unions and other attempts to raise the standard of living of working people. The 1950s were a period of enormous global growth, rising income levels, and a time of adaptation to the new world order after World War II. This decade was possibly the last before the golden age of affluence. Once that hit, all hell broke loose; we ironically became more comfortable than ever and simultaneously more pissed off than ever.

The sixties have a special place in western society. For the first time in history, an entire generation was able to not worry about the future, which allowed them to focus more on finding out just what the world looked like with enough drugs in their system to tranquilize a horse.

It was a heady time for young people, with all the old squares working like dogs at ever higher rates of pay, using their income to provide their children with all the things they never had in their youth (like LSD). The drugs were at least good for music, and the world was brought together in consciousness by global entertaining phenomenon like rock music and new-fangled television sets.

We can sum up the entire decade in a few essentials: Rock and roll, cars, global travel, drugs, Russian expansionism, the Cold War, more drugs, Vietnam, Vietnam protests, military industrial complex protests, and some more pills of unknown origin. During all this, China was still awakening to the modern world, and the Middle East was a poker game where the multinational oil companies were the players and mega-oil reservoirs were the chips.

From a protest perspective, all the antics of the sixties first led to sort of soft protests against working for The Man, and challenging the industrial complex that had come to define western economies. It is true that many protests centered on race as well, a situation which had been simmering for decades. Those protests parted ways with the mass 1960s movement that wanted to take on more global fights like the war machine. This soft protest then morphed into something else with the catalyst of the Vietnam War, a hugely unpopular war as wars go, which from some angles appeared to be designed to support military spending and innovation. Many hippies were extremely dismayed to see big corporations apparently thriving from the war (when they could see anything at all).

As the protest business morphed and grew in the 1960s, a subtle undertone emerged with ever-more force. The protest movement had origins as a social justice movement, where battles and strikes were often over working conditions or unionization. An undercurrent of these movements was a hatred, or at least a deep mistrust, of corporations, which also went hand in hand with the Russian vision of promoting Marxism. North American university professors enthusiastically endorsed the tenets of state ownership and control, which directly impacted the attitudes and values of youth about to head out into the world and begin their adult lives as post-graduate ex-drug fiends.

This suspicion of corporations may well have been warranted, but I don't know because I wasn't there, and whoever tells you which side was to blame is probably lying. We can sum up several decades of "he said/she said" with the observation that no matter what the working conditions were,

MODERN PROTESTING: AMATEUR HOUR NO LONGER

most jobs were better than starving to death, and at the same time there was no excuse for abandoning basic principles of human dignity in the pursuit of ever growing profit (It feels good to be able to piss off the two pillars of the political establishment in one sentence. Now that's a protest.). What matters in this context is that the protest movement was anti-establishment, and that meant to a certain extent, anti-business.

Thus began the modern protest era, where social causes and a certain anti-business mentality took over. Protesting against industry and industry/government partnerships/collusion became sort of fun, because there was no downside. Western governments would rarely throw people in jail (though of course a few were killed here and there at more vigorous protests, but that just made the whole pastime kind of spicy and stimulating), or if they did get thrown in jail it was the "hey-I-got-thrown-in-jail-too!" kind of bonding experience that had no repercussions (at least for white people).

Regardless, some good has definitely come from the protest movement, such as when it got behind worthwhile causes like the fight against racism or homosexuality (my favourite protest sign, seen at a gay rights rally: "If God hates us, why are we so cute?") or joined forces to fight a multinational corporation in order to, for example, bring back the McRib sandwich. Protests were also valuable in raising awareness of environmental debauchery that was happening out of sight of the freeway, because the limited communication tools of the day did not facilitate examination of that sort of thing. When forests were being clear-cut way out in the middle of nowhere, it was good that people were made aware of this before they had their own drones to check it out.

The list goes on. Many protests have been worthwhile in helping upgrade workplace safety or other worker-rights issues. Many people have reason to be grateful that those protests did occur. Unfortunately, that is not universally the case.

PROTESTING AND SOCIAL MEDIA

An interesting aspect of protesting is that it is usually one-sided. That is, one camp protests, and the target of the protests looks like they've been hit in the face with a shovel. There are exceptions, like with guns and gays (that would make a great name for a band) where each side can muster up enough people to form a decent mob. For business-related protests, though, it is a thoroughly one-sided affair.

Pipeline companies are the most recent and vivid example. What makes this situation peculiar is the fact that the object of the protest was until recently viewed as the most boring, stable, and inert form of business imaginable. There is no joy in operating a pipeline, no glory or admiration from anywhere, just as there is none for furnace ductwork engineers. This isn't a slight against either profession, it is simply a fact that some professions and asset depreciation classes simply do not ignite the public's imagination. That is, until they ignite the public's things, or spill oil all over them.

At any rate, pipeline companies became the target of protesters in recent years, and the phenomenon left them thoroughly unnerved and on their heels. The protest business aggressively seeks out the weak spots and amplifies messages that are difficult to calmly rebut. When protesters targeted the Keystone XL pipeline, they attacked on all sorts of fronts, from the dangers of climate change to pictures of oil-soaked ducks. For an industry used to decades of calm and quiet underground operations, an effective response was extremely difficult.

In the current environment, rather than face protests, we now see companies react almost violently in response to incidents that may damage the public trust. Companies almost panic when something bad happens, and defecate a torrent of apologies and sensitivity training programs when caught by some off-hand video, like when two black men were arrested in a Starbucks in Philadelphia in April 2018 while waiting for a friend. Fifty years ago, this would not have made the news for several reasons, one being that the episode would not have been filmed and immediately disseminated around the world, and secondly because at that time these antics were a way of life for black people. Starbucks, in response, displayed the latest in cutting edge PR tactics: it closed 8,000 outlets for an afternoon for some sort of sensitivity training or something (which was an unfortunate and wasteful disruption of business; a thirty-second, universally-delivered message along the lines of "What the f___ is the matter with you?" could have done the same thing).

In the fossil fuels business, it's not that easy. Starbucks, while having a big problem on their hands, had only to show the public that they understood what happened was wrong, and that they were doing something about it. For fossil fuel businesses, such as pipeline companies, how could they defend against messages insinuating that fossil fuels were going to destroy our children's future by causing global warming? To say that pipelines are safe is well and good and true, but the sophisticated strategies and tactics of protesters shaped the battlefield so that such messages only served to worsen the problem for the industry.

What has put older established companies, including most of the petroleum business, on their heels has been the rise of social media. In the next chapter, we'll look at how the new world order of instantaneous and easily-manipulated communication has changed the landscape, with one side of the fossil fuel debate taking to it like a duck to water, and the other so far behind the learning curve that its efforts to get there are simultaneously sad, comical, and potentially suicidal.

11.
Out of touch: 1950ˢ PR strategies run head first into social media guerilla warfare

BIG, COMFORTABLE OIL

A funny thing happens when you are lord of the universe. I am not of course referring to myself; my kingdom is about the size of a modern closet. No, that is a reference to the oligarchs of the petroleum world, the global energy titans. Imagine how they view the world of fossil fuels. We can be assured their view is somewhat different from ours. For us, if in a philosophical mood, or at the least experiencing the philosophy that arises when filling our vehicles with expensive fuel, we may reflect on just how addicted we are to fossil fuels. In the Versailles-like caverns that are global petroleum executive offices, they *know* how addicted you are.

They've known it for a long time, ever since automobiles took over the planet and coal became a second-class citizen relative to the juggernaut that petroleum was becoming.

There was no competition in terms of other fuels, and the world's reserves were there for the taking because the oil giants had the exclusive know-how to get them out. Foreign governments welcomed them, brushing peasants aside or neatly putting them in piles as required in the quest to access petro-wealth. Other competitors were no big deal either, they simply gobbled each other up, with the redundant executives retiring and laying on cubic acres of money, or if they felt like it starting a new oil company just for fun.

It would be truly weird indeed *not* to develop some sort of superiority complex, given that the world's appetite for petroleum continued to rise relentlessly, and that you were one of a relative handful that knew how to do anything about it. What could possibly go wrong?

LOBBY? WHAT LOBBY? JUST DO WHAT WE TELL YOU OR TEN THOUSAND JOBS DISAPPEAR

There is a commonly held belief that oil companies bend governments to their will through an army of lobbyists. That is somewhat true, in the sense that big oil companies had massive presences lingering around the world's more powerful governments. But it is a bit of a misnomer to think that lobbying was their game; there is often no need to stoop to that level.

Lobbying is—per the Yoda of our times, Wikipedia—the act of attempting to influence the actions, policies, or decisions of officials, usually legislators or regulatory agencies. That seems a fair enough definition, and seems completely consistent with what we associate the icky profession with: whispering quietly into the ear of legislators, promising this or that, pandering to vanity, and whisking away to Vegas the hapless victim to enjoy the city's best assets in the best hotel rooms that may or may not have hidden cameras in the plants. Oh wait, that would be illegal, so probably never happened at all, ever, not once.

It is a mistake to lump the oil barons in with that lowly crowd; petroleum's statesmen surely looked down their noses at such boorish behavior. They could do so, because they had a hammer of the gods that the lobbyists could only dream of. They controlled the lifeblood of the world's growing economies. Actually, they *were* the lifeblood.

Governments were also well aware of what petroleum did for their standard of living: for consuming nations, it was a godsend because it enabled development, and for producing nations it was heaven on earth as countries with relatively small populations could find themselves wallowing in money up to their armpits through petroleum development.

Big oil, therefore, became not just a bunch of grubby lobbyists scurrying around the halls of power along the baseboards and biting each other, but the guys who strode down the halls of power right alongside heads of state. Those were the channels of communication they lived in; they did not speak to the public or anyone else in efforts to drum up support for whatever they were up to. Why talk to the media when you can talk to the president?

PR STRATEGIES RUN HEAD FIRST INTO SOCIAL MEDIA GUERILLA WARFARE

It was no coincidence that Donald Trump, upon becoming president, selected Rex Tillerson as his Secretary of State. As head of ExxonMobil, it was felt, correctly, that Tillerson had developed the skill set to step right into one of the world's most preeminent statesman roles. Tillerson had more experience dealing with global heads of state, including a hyper-finicky, recalcitrant, and calculating Russia, than most in Trump's inner circle. In fact, Tillerson's qualities as a statesman seemed to be on a higher level than his own boss, whose diplomatic skills are presently equivalent to what Tillerson's were when he was three. Of course, this didn't last; Trump got rid of Tillerson because he was too statesman-like. These are strange times indeed.

The concept of "lobbying" then hardly applied to oil companies in the common sense of the word. The big players in the oil business were far beyond that demeaning term. They could pretty much do what they wanted, or threaten to withhold development capital, or simply threaten to leave.

This influence did have its downside, as a handful of countries that owned a majority of the world's reserves began to realize that that power that was available to them, and then some. Small oil producing nations took notice of how the world listened to energy producers and decided that that looked like fun. Plus, they had grown weary of having their fate dictated by big oil. The overbearing control exerted by the large multinationals led to the creation of OPEC, the Organization of Petroleum Exporting Countries. Owners of much of the world's reserves, these nations finally decided they'd had enough and kicked the supermajors to the curb.

Even then, the world's supermajors had a lock on things. OPEC may have come to own much of the world's production, but that oil eventually had to go to refineries and distribution systems largely controlled by the supermajors. Life continued to be good, and the concept of global goodwill remained sort of an alien concept in the halls of petroleum power.

It's not hard to see how this entrenched position of power that the supermajors enjoyed led to a certain complacency and comfort level. Even OPEC's manic manipulation of oil prices worked out in their favor: every time OPEC rattled chains to raise prices, the profits of the supermajors soared. Oooh, hurt me. If prices fell, demand grew even faster, and the supermajors controlled the refineries and distribution channels, so again things worked out not too badly at all.

That the world was starting to move in a different direction was not a major concern. Every new development in technology, and every new convenience that the world became accustomed to, simply meant more energy. More energy meant more fossil fuels. Nuclear power had been discredited, and there really was no other substitute for hydrocarbons. Oil prices went up, they went down, it didn't matter; the supermajors owned the refineries too. Money flowed in from somewhere, and demand kept going up.

The biggest task for big oil was managing geopolitics, which is the reason Trump brought in Rex Tillerson to handle the delicate art of international negotiations (Trump was clever enough to bring Tillerson in, but wasn't clever enough to overcome his natural instinct to pummel Tillerson into the ground). Trump was right, initially: a certain smooth skill set is required for successful international relations. You think you have trouble dealing with a crazy neighbor whose dog craps on your lawn. Imagine heading into a military dictatorship to try and negotiate a long-term oil exploration and production agreement. That's what the world had become for big oil for much of the twentieth century.

It wasn't hard though, because big oil had a few things going for it. One was enormous wealth, and the other was the support of the US military machine. It was primarily an American game to circle the globe ensuring a reliable source of petroleum, and "reliable" is sort of a euphemism for "sign right here or we'll wipe you out."

The point is that the petroleum industry in general—the old, overarching industry—viewed the entire oil business as a tactical process that could count on government and military support to achieve the objective of securing oil supplies for the western world. In the 1900s, these things were not easy, but easily demarcated. The enemy was clearly visible, and usually bad by western standards (godless communists, corrupt dictators), and the battle grounds were clearly visible.

When the world's governments pander to you, and the world is hopelessly addicted to your product, the question arises again - what could go wrong?

WHAT'S A FACE BOOK? NEVERMIND, WE'VE GOT A BUSINESS TO RUN

Times have changed remarkably for fossil fuels with respect to their stature in the world. The military-industrial complex that defined the sixties, seventies, and eighties has changed significantly. The world now moves at warp speed, information flows everywhere instantly, and the most feared enemies of the petroleum industry walk among us. They don't stand on the other side of a physical barrier; they are not obvious solely by their uniforms or by their funny foreign talk.

The old ways of international energy policy don't work anymore either, not just on the production side, but also on the consumption side. A new world order of foes has arisen, one that is, for the first time in petroleum's history, seriously challenging its de facto place as the absolute necessity of modern life. It *is* still the absolute necessity of modern life, but there is a colossal change that has happened in the tone of global discourse with respect to energy. You may have noticed that.

In the previous century, governments aided and abetted fossil fuel development because it was obviously the engine of economic growth. The US had showed the world a new way to live, and Americans drove cool cars all over the place and life was like one big beer commercial party, and the rest of the world wanted in too. Cheap energy was the key to that, and the search for, and access to, reasonably priced petroleum was paramount for every nation on earth that wanted to see what it felt like to live in a society that could invent and afford, for example, monster trucks and hot dog eating contests.

This was the frame of reference for global petroleum companies: that the demand for petroleum would always be there and would always grow. There was no need for public relations to speak of. Sure, they all had their departments that had to deal with regional issues and talk about how much they cared about communities, but from a macro perspective, it wasn't really serious. When governments have your back and fight to bring you in, an oil spill or local land claim or pipeline right of way is as much an obstacle as a clay pigeon is to a shotgun. Western media would occasionally feature stories about some developing-world oil-related embarrassment or injustice, and the better parts of society would be outraged, but the essential response was a shrug and a what-are-you-gonna-do reaction. We all need oil just like we need hamburger. How that beef gets on your plate isn't all that pretty, and we just don't talk about that either, but we still eat the beef.

CLIMATE CHANGE BOMBSHELL

In the past decade or two, the semi-cozy relationship between big oil and big government has become strained, to say the least, because big government is cheating. It no longer just pays attention to big oil's heavy lipstick, overabundant perfume, and outfits it should not have been wearing for twenty years. It's now answering to the granola-eating athletic type. That is a sea change of monumental proportions.

Big oil institutions are like any other big system, they do not change easily. They can't. Imagine management structures and hierarchies put in place over the better part of a century, administered by a workforce of up to a hundred thousand people. It can take an international effort involving a dozen countries, five legal groups, countless business groups, and three years to change the corporate brand of in-house coffee. I have been there, and it is like watching rocks erode.

Imagine then changing the entire PR mentality of these companies from one that is completely relaxed and unafraid of waging geopolitical battles because governments support them, to one that has to deal with antagonistic governments.

Now, imagine doing that based on a PR structure that has been in place for decades, and that really isn't a PR structure at all; one that has comprehensively ignored social media because, well, it was irrelevant, but now must pick up those tools and find out how to use them.

Now, imagine going into social media battles with an opponent, the climate change/environmental movement, which has not only mastered these techniques but leverages them in truly spectacular fashion.

Now, imagine that those climate change/environmental groups are clever enough to make the climate change debate an ethical and social welfare issue, while big oil is trying to figure out how to get a tweet out.

Add all that together, and you have an explanation for the tactical bloodbath we get to see every day in the news.

PR STRATEGIES RUN HEAD FIRST INTO SOCIAL MEDIA GUERILLA WARFARE

It's not just big oil either; the entire energy complex has been encircled. The view from inside the energy business has always been more or less the same, no matter the size of the company. The view is that common sense will prevail and people will realize that those finding and producing fossil fuels are key to their survival. Finding petroleum is actually pretty hard, and since the world absolutely and unequivocally needs it in vast and growing quantities, at the end of the day the world will get out of the way and let them find and produce it.

That view was yet another tactical error. The environmental/climate change opponents of fossil fuels wisely seized on a fundamental human trait that we looked at a few chapters ago: people take things for granted over time. People don't get ecstatic when they walk into a heated building. The same goes for air-conditioned buildings. When people fill their car with gasoline, they are far more likely to curse the price of fuel rather than be grateful it is there.

12.
Key messaging warfare

LEARNING THE MESSAGING TRADE FROM TURTLES

About a decade ago, through some strange circumstances, I found myself as a spokesperson for an energy infrastructure company. I'd not had any exposure to corporate communications or PR at all, but had been hired to write news releases and other enthralling corporate material. The media aspect of the job was almost nonexistent, so a lack of experience in that field was not much of a problem. At the time, energy infrastructure companies did not like media, or media exposure, unless it is on the business channel and the discussion is about dividends. Being a spokesperson was not the most active part of the job.

Nevertheless, there I was, in charge of the whole public face of the company, because under the corporate communications umbrella fell the function of company spokesman. Given that the business occupied the less-intoxicating end of the corporate spectrum, "media training" equated to "disaster planning," because that was the most likely scenario in which such companies envision themselves being in the media. For an energy infrastructure company, the disaster could be a spill, a chemical fire, or industrial accident of some sort.

Once ensconced in my new role, I set about learning the fine art of dealing with the media. This involved some important training, as anyone who has ever faced a taped interview will know. The media trainer was doing an excellent job of preparing a total neophyte like myself for a potential life in front of a camera, with (what felt like) hours of on-camera practice (actually, only uncomfortably awkward minutes, as the video evidence horrifyingly confirmed).

Media training was crucial to the job, because dealing with the media is not a natural thing to do. It became apparent that the training was composed of two parts: one was to become comfortable in front of a camera or before the media, and the other was to become comfortable answering questions. I thought that was the order of importance, because the best spokespeople I'd seen in front of a camera were extremely comfortable and answered every question with ease. I was quite wrong. Being comfortable in front of a camera was insignificant compared to getting the answers right, and that was a far different skill than I'd imagined.

The lessons centered on the usage of "key messages," a phrase that had little meaning to me before taking on this assignment. Once in the know, however, the phrase comes to mean the communications equivalent of how to deal with a medical "code red" or nuclear power "meltdown in Sector 7" type of mental defibrillator. Key messages are what you reach for in event of an emergency.

As it turns out, the first and only rule that I was to keep in my subordinated little head was this: here are your four key corporate messages, and if you f__ing dare stray from those, you are dead meat. This wasn't a corporate threat; it was implied by the media trainer that I would be eaten alive by the media if I went free form.

In the job of corporate spokesperson when dealing with a crisis, there is nothing more critical than knowing the corporate key messages and hanging onto them for dear life like a piece of flotsam after an oceanic plane crash. For communications rookies such as myself, however, it was a bit of a tough slog for the poor instructor.

"Suppose you're manning the barricade on a road that leads to one of your facilities where there has been an explosion and possible fatalities. The media approach you with questions about the incident. What are you going to say when they ask you about what happened?"

"I'd give them an explanation of what happened."

"Wrong. You don't know anything for certain yet. What would you say instead?"

"We have no comment at this time?"

"Wrong. You never say you have no comment. You'd tell them what you know and always, always, circle back to the key messages. In your instance, the key message is that safety is our number one priority."

"How could I say that if I didn't know what happened?"

"Then you'd say an investigation is under way, and that safety is our top priority."

"But what if they could see a fireball over my shoulder?"

"Safety is our top priority."

"What if that made no sense in the context of the incident? What if ambulances were screaming by as we spoke?"

"Safety is our top priority."

"But no one would believe that if they were watching me live on TV with billowing black smoke coming from right behind my head and obvious injuries."

"Safety is our top priority."

"What if they keep peppering me with questions about what is so obviously going on?"

"They won't."

"But they can see it."

"If they aren't getting anything out of you, they'll quit soon enough, but only if you never deviate from the script. Safety is our top priority. If you deviate, you're dead in the water. If you don't believe me, let's go look at your practice interviews again. Safety is our top priority."

The tactics mentioned above are, of course, not wrong; that is what's taught in modern media training, and those lessons are taught for a reason. The problem is that they are symbolic of an industry run by former masters of the universe who are seeing their world slip away from them, and they don't really know how to deal with it. As with everything else, they take the most conservative way imaginable, because it's viewed as the safe way. Withdraw your head and limbs until the menace goes away, and re-emerge to fight another day.

Contrast their attitude with Elon Musk's and the way he runs Tesla. On a recent 2018 conference call, he cut off several questioners saying their questions were boring, then devoted twenty minutes talking to some guy that called in who had his own website "because he asked interesting questions." He regularly takes to Twitter to convey all sorts of messages. He's not scared to say anything, he tackles challenges as they arise in the media, and he reaches out to correct what he feels is disinformation. He does it in a style that people understand and relate to. They may not always agree, but they always notice. When he oversteps the bounds of acceptability, he apologizes and moves on, or, at the very least, settles up with the Securities and Exchange Commission.

In the end, the Tesla Twitter example is a microcosm of the situation that envelops the energy world. As an example, no one likes ExxonMobil, and lots like Tesla. You could say that it's not a fair comparison because Tesla is modern, cutting edge, and wants to be liked. They engage with media, markets, and people. They seek to connect with the world around them. ExxonMobil may want to also, but we can't tell because whatever heart is beating behind the pile of lawyers is unobservable.

There is of course a reason why ExxonMobil operates the way it does. Most huge established corporations develop along these lines. For the right situation, the communications strategies of ExxonMobil are valuable. There is a reason that ExxonMobil's top executive Rex Tillerson slid easily from the company's big chair straight across to dealing with Vladimir Putin and that level of global grandiosity in such a seamless manner. But that old structure is useful mostly when dealing with high-level diplomatic sessions, where months are spent planning everything from the menu to the colour of ties. That mentality and structure is perfectly useless in the social media trenches.

When Twitter started to gain popularity and widespread acceptance, the company for which I was spokesperson took one look at it and started laughing. In those hallways, public communication was a sacred process that was controlled with absolute authority, and would never be done whimsically or trendily. A senior management group made it clear in no uncertain terms that the company would never be tweeting anything, ever. Not only was the platform not embraced, it was openly mocked. And this was less than ten years ago.

The old-school mentality that considers social media platforms like Twitter as irrelevant spells trouble in, for example, the event of a major incident. There is no contest in the fight for public opinion, because these days it is fought in social media at lightning speed. Savvy opponents pummel slow-moving behemoths into the ground like tent pegs before the first corporate news release had been drafted, circulated, edited, recirculated, approved, uploaded, proof read, and released. While the key messages of "safety is our number one priority" and "we will provide more information as it comes available" are hitting the newswire, some farmer with a cellphone will have uploaded live video of oil streaming into a river or pictures of a coated duck that will have already gone around the world three times.

Global energy titans are taking tiny steps into the world of social media, but the timidity and blandness of their approach interests almost no one. As I write this, it's mid-month, on the fifteenth day. Thus far in the month, Tesla has published seven tweets, with an average "like" count of over 18,000. ExxonMobil's Twitter feed for the same time frame has ten tweets, and average number of likes is forty-six. ExxonMobil has a presence around the world and has 70,000 employees, of whom it appears forty-six like what the company is up to. To be fair, that's the average; in this small sample, as many as several hundred liked the company. It gets worse: that number includes the whole planet, and it is probable that at least some non-employee approves of ExxonMobil and would be enthused by their tweets. So that leaves a handful of actual employees, a number you could fit in a Volkswagen, that are willing, able, or interested enough to bother clicking the like button on Twitter on a consistent basis. The rest of the world is not engaged.

EVEN THE GOOD STUFF THEY GET WRONG

The conservative old-school strategies have further drawbacks, because when it comes time to actually try to get a message across, it's virtually impossible to do it effectively. Decades of messaging conservatism mean that the default habits tend to embrace silence rather than risk a misstep.

Sometimes these corporate behemoths refuse to try at all, for fear of… well, I don't know what they're afraid of. By way of example, a few years ago some unusual weather events happened in western Canada that culminated in severe regional flooding. The company for whom I was spokesperson owned some assets that *could have* been damaged by a flood, if it did indeed occur at certain sites, and the company took no chances. It shut down the flows in the area and removed all petroleum, just in case the flood did happen. The company had a thorough knowledge of exactly where, in the event of the flood, the potential for damage was highest, and which particular infrastructure *might* be impacted. If it did happen at that location, it was assumed that the flooding would be large enough to cause extensive damage, and that the products may have been released into a waterway. In other words, the precautionary measures were exemplary and would have been, if mentioned to the public, both exonerating and jaw-droppingly impressive. No one would have guessed in a hundred years that a company like this would have been so well prepared that it knew in advance which of thousands of locations would be at risk, and so concerned about the possibility of an incident that all oil was removed and all activity halted, just in case. For a public led to believe that petroleum transporters are blasé about oil spills, this one example would have silenced many critics.

Bizarrely, us communications people were not allowed to mention this brilliant act in public. Efforts to issue a news release about these masterful safety measures were quickly quashed; instead, a press release was issued, announcing that the company was monitoring the flood situation carefully. No one knew what on earth that meant, because we wouldn't tell them. This impressive safety-precaution anecdote never made it past the first executive edit of that year's annual report. I tried a watered-down version, and it too was surgically removed. There was no mention whatsoever about the measures taken other than to offer some platitude that the year had a highly satisfactory safety record and that the flooding was successfully managed.

Living through these events gives one the ability to state with a high degree of confidence that such entities haven't the slightest idea what war they're in. They surely know they're in one, all right—the evidence is obvious and unavoidable from any news source—but they don't know what to do about it. It's as if they're stuck in 1950 when the enemy spoke a foreign language, dressed differently, and stood on the other side of a wall with their rows of weapons. Ah, the good old days, life was so simple.

The war being waged today is fought from some nondescript building by social media experts who have fifty Twitter accounts set up to echo their messages and get noticed, and by the time anyone pays attention to whether these messages are true or not the next wave has been issued and the earlier forgotten. The offices of these virtual warriors could be around the block from big oil HQ or on the other side of the world. This all happens while the top-floor executives are ten steps behind what is actually happening on the ground, and well before approval for any sort of corporate communication is granted. It's guerilla warfare—modern ninjas equipped with all the latest weapons pitted against knights who need five assistants to help put on their armour and get on the horse.

THE EVIL NECESSITY THAT IS KEY MESSAGING

The earlier account of the single-minded focus on simplistic key messages is not necessarily an indictment of the practice. It is a sound strategy in today's communications world, particularly when the mic was handed to an ignorant buffoon like myself. Reporters are quite skilled at leading interviewees off down some garden path, then planting a rake in the grass to see how the story changes when the handle hits the face at half the speed of sound.

Key messaging is critical and a part of the landscape. It's a dirty job but someone has to do it. The media trainer was simply teaching a survival strategy. This is the way the modern world works: adapt or die. That might be a bit melodramatic. It's more like adapt or be a sweaty stammering mess on live TV, uttering impromptu theories that would have corporate lawyers cringing like beaten dogs.

The problem with the key messaging that was driven into my head is that the content was hopelessly naïve and laughably ineffective. Correction: it is effective in the shortest of timelines and at the shortest of radii, being designed to allow a corporate spokesperson to gracefully exit a potentially dangerous media situation. It is completely ineffective and even harmful with respect to the wider battle. Take pipelines for example, and the textbook spokesperson-chant that pipelines are safe. They are safe, but it doesn't matter when it's the pipeline spokesperson assuring us that. In isolation, say if you landed on a desert island whose inhabitants had never heard of a pipeline before, and you showed them one, and moved some fluid through it, and then declared "pipelines are safe," you might get some contemplative nods and shrugs that said, "I can't see why they wouldn't be, based on what I just saw."

What the poor corporate PR strategists didn't, or don't, seem to realize is the power and speed of social media to trump those messages and turn them into a symbol of how out of touch the petroleum business was and is. All an environmental group had to do was tweet a quote from a "pipelines are safe and safety is our number one priority" corporate PR piece, and attach it to a picture of a bird covered in oil or oil floating on a Michigan creek, and that message will go viral while the corporate press release will be unread, and the company would have been better off if it had never opened its mouth at all. The environmental group didn't even have to say anything.

I once had a chat with a member of a group that was working on messaging for the pipeline industry, and he relayed a story about a focus group they'd organized to evaluate strategic key messages. One message read something like "99.99 percent of products move safely through pipelines." The focus group responded in two ways: one sub-group thought that the message was terrible because it was a lie, and the other bunch thought the message was terrible because it implied that 0.01 percent was acceptable, meaning that for a pipeline system moving a million barrels per day, a hundred barrels per day of spills was acceptable. That small percentage equates to 36,500 barrels per year on a million barrel per day pipeline, and if anyone thought that was acceptable, they needed their head examined.

It's not that pipelines are not safe, or that safety is not the number one priority. Those are undoubtedly true. But it's also true that those messages are not just knives to a gunfight but like getting a tattoo of a target on your face before heading to the gunfight. They are fodder for social media experts.

This chapter is not to single out big oil, or any oil, as being at the forefront of pathetic PR policies. Their PR is the best they can hire, or put another way, the best they hire that will work within their structure. What is notable for the purposes of this conversation is how bad the oil companies are at the current game, and how being a master of the universe with endless available cash does so little good in the current media world.

SO WHY DON'T THEY CHANGE? IT'S THE SYSTEMS, AGAIN

Did you ever watch the first rock bands to emerge from the former Soviet Union? Many were quite endearing in their sincerity, even if their overall shtick made you cringe. They never grew up in the rock and roll culture, and only caught it when it became available. They had zeal and they'd seen the world's coolest, and that's where they were aiming. But there's something elemental and useful about having time to understand fully the new world you're aspiring to. U2's first gig was probably embarrassing, but they were thirteen so you'd cut them some slack. They became masters of their game over a decade, at minimum. Thinking you can go from zero to Bono in four months makes for painful entertainment.

We see the same thing when corporations or their spokes-groups try to leap into the social media fray without having a history of concerted and effective public engagement. It looks so easy; hire a few millennials and we'll be up to speed by Q3. Fire up the Twitter.

Oh, but wait, before it gets to that, the senior executives have to have their say, based on decades of conservative corporate communications management. We'll need a single writer for consistency, or better yet a committee to ensure not only consistency but to remove any element of personality. Second, anything we're going to say has to be consistent with corporate policy, and it has to be consistent with everything we've said before. So yes, a committee would be necessary to ensure continuity. Then we'll need to run the messages by the business units to make sure they're factually accurate. Then, just to show they have a sense of humor, someone around the board table asks, "What do you all think if we do this without consulting legal?" That one brought the house down. Oh my, the mirth. The board hasn't laughed like that since… well, they've never really laughed before.

You can imagine what the outcome of this bureaucratic routine is. It is extraordinarily like the human body, where an exciting and diverse array of foods and flavors and textures go in one end, and out the other end comes a certain unfortunately homogenous product. The food in this analogy would be the thoughts and observations and enthusiasm of the younger generation and the business unit leaders, who actually do things and that are worth hearing about in the real world. As for the byproduct in this biological analogy, let's just say the corporate approvals committee would be the equivalent of the colon.

There really isn't an easy way around this phenomenon. There is no way on earth that ExxonMobil could start acting like Tesla in any reasonable timeframe because of its inherent systems. As you will recall from the last dozen examples, systems are hard to change. This fact does not go unnoticed by the modern media machine and its A-list players.

PR HAS ALWAYS BEEN CREEPY. NOW IT'S REALLY F___ING CREEPY. HERE'S WHAT THE PROS ARE UP TO.

Messing with the media to influence news streams is nothing new. In the old days of print media, hucksters attempted to get into the papers by creating sensationalistic stories and claims, sometimes succeeding and sometimes not. The path to doing so was obvious though: one had to deal with a reporter and convince said reporter that the story was genuine and worthy of inclusion in the newspaper. That seldom worked with cagey press people, who were more interested in genuine news than being used.

Consider what a "breaking story" meant fifty or sixty years ago. Wherever something newsworthy happened, word of the event had to reach a news organization, then a reporter had to be dispatched to the site, then investigate, then write it up, then file that report with the paper or TV station, which would air that night or the next day. From the time of the event to widespread dissemination of anything substantial could take the better part of a week.

Today, as everyone knows, it all happens almost instantaneously. A cell phone can capture images of a spill or incident before even the owners/operators know, and upload it from almost anywhere in the world.

The real story will follow at some point if the event is significant enough, but by then it will be a reactive story, trying to catch up with the images and messages the world has already received, analyzed, and judged. A corporate key message therefore becomes irrelevant when the first pictures to be presented unto the world carry not the message that the company wants, but the byline of whatever that Twitter user wanted it to be. That becomes the identity.

What's most unsettling today is how cutting-edge social media experts have seized these new tools to upend any conventional sense of normality in the news. Quality, in the massive daily news flow, means little. What matters are views, clicks, and attention. The number of viewers has always been important, but that had an entirely different meaning when people bought newspapers or had a choice between only a few TV channels. Now, the web is the source of all information, and to turn to it for news is to have a hundred million "look at me" icons bouncing up and down trying to get your attention.

Which one of these you eventually look at is not random. You are most likely to be confronted with those that are most popular, or as we say now, "trending." Those are the stories that will form your opinions of the news and the world. Those views are for sale, and as it turns out a sub-industry has developed that will help the highest bidder manipulate the media (and the masses) like a hypnotist makes a middle-aged man act like a chicken.

THE NEW BLACK OPS

At the time of this writing, a once high-flying company called Cambridge Analytica (CA) is packing up its junk and heading for the elevators, forced into bankruptcy. CA was one of the new life forms that arose from the petri dish that is the World Wide Web. The food on which these life forms thrived is personal data that is available, free or otherwise, on the net.

Their world begins with companies like Facebook and the like, places where people happily and freely share with the world pretty much anything.

Businesses like CA are pioneers of target marketing, such as promoting a particular politician, based on the demographic information they possess. This is what brought CA down. The firm had apparently stepped over some sort of line by sucking data out of Facebook and selling the information (or use thereof) to the highest bidder in the US presidential election. It's a pretty greasy business, but so is politics, and not much can happen in that realm that would raise the average eyebrow.

On the other hand, some new masters of social media have risen up out of the muck that should really be of interest. This new breed doesn't just gather information on you to try to get you to do something, it uses tools of deception to control the media stream.

An article appeared on the Trumplandia website, a site created by the Columbia University Graduate School of Journalism, about an investigation of modern PR techniques on the web.[1] It profiled several "comms strategists" who push the envelope of distortion. One woman named Sally Albright, who in true mercenary style worked for both Hilary Clinton and Newt

Gingrich, was documented to have created dozens of Twitter accounts that appeared real, with facial images lifted randomly from web pages to give the impression that real people were behind the accounts. Ms. Albright would, depending on what "campaign" she was working on, tweet, retweet, and probably re-retweet hundreds of messages, giving the impression that certain conclusions of innuendo or rumors were true and widely believed. She was a master of trending.

In another more explicit example, the Trumplandia article outlined how a native group had been working for several years to get the Washington Redskins to change their name to something more politically correct. Their campaign had lost momentum, until one of the group's founders was approached by a social media whiz kid named Jordan Daniel, who has done a lot of work for worthwhile causes around women's rights, but is also a "culture jammer." This term describes a new breed of manipulator that is at the forefront of destroying whatever trust remains in the media. Daniel offered to help the native group sharpen their tactics by creating a phony news event about how the team had changed their name, including the creation of a fake website. The media lapped up the false story and it hit many of the mainstream publications. Quickly, the native band owned up to the mischief because many natives felt hurt and betrayed by such deviousness. The organizers of the scheme were completely unapologetic because it worked so well.

Sleazy? You be the judge. No wait, I can handle this one. Yes, as sleazy as a pimp. Effective? Undoubtedly. To make the waters murky though, what if this data-mining and media manipulation had been done for a good cause?

What's ultimately disturbing then is not the incidents or the revolting people behind them, but that this is what it's come down to. This is how one succeeds. Some of these pioneers, such as the women's rights advocate mentioned above, would argue that the existing media stream is devoid of standards, decency, and honesty. That may well be true, but it is a weak argument indeed to think that one is doing good work by acting even worse.

EVERYTHIING OLD IS NEW AGAIN

As with everything else we've looked at, it's the established systems that have gotten us to this place, ones that developed over decades. The system is simply the means of disseminating messages to the public. It used to be newspapers, then radio, then television, now the internet. Each was faster than the one before and more subject to manipulation.

At one point, the manipulation was more blatantly practiced by the corporate community. They could own the communication channels, or pay for all the ads, which is always a big help in twisting messages. Corporate manipulation tended not to be so much message twisting though as corporate development, at any cost. Cigarettes were the most visible example, with ads plastered everywhere including bizarre paradoxical pairings such as the du Maurier Open Canadian pro tennis tournament, du Maurier being a popular brand of Canadian smokes during the day. (I yearned to see a match between two players each with a cigarette dangling between their lips, who's your daddy sort of thing, but I was never fortunate enough to catch one.)

With the rise of the internet and all the free social messaging services, it's a free for all and media manipulation goes to the quickest and the sneakiest. Big oil may be wise and rich, but it is neither quick nor sneaky.

CULTURE JAMMING PIONEERS WILL RUE THE DAY THEY EVER STARTED THIS

Right now, in the battle for public opinion, big oil and other similar big fossil-fuels are on their heels, hopelessly outgunned and outmaneuvered by the nimble upstarts. The opponents of fossil fuels have all bases covered, making construction of new infrastructure extremely difficult, and blockading or demonizing anything that is linked to coal, oil, or natural gas.

But the story doesn't end there, not by a long shot. Recall that the global energy mega-system is, like all others, incredibly difficult to change. Even pro-environmental forecasts envision massive fossil fuel usage on a global basis for the next 20-30 years.

The demonization of fossil fuels is making for a global population of hypocrites, and unless a significant proportion stop using fossil fuels in the near future, the impact of the extreme anti-fossil-fuel messaging will wear off. Not that the environmental movement will die off, or that people will stop the change towards green-ness, but that the attacks will lose meaning, like accusing people of disobeying the rule of law by jaywalking. They will simply agree that they are lawbreakers when it suits them, because giving up jaywalking is not going to happen. Similarly, people who are standing at the pumps filling their car with gas or enjoying a nice warm room in the dead of winter will at some point become immune to the messaging that fossil fuels are deadly.

Much worse for the game though is if big oil gets its act together. Remember that fundamentally, the internet is a commercial venture, and eventually money is what counts. Presently, the most aggressive social media mercenaries are on the environmental side, and receive substantial funding from various deep-pocketed organizations. When it turns into a bidding war though, would you bet against the likes of ExxonMobil? How about ten of them?

It has long been a sad phenomenon when social institutions that serve a useful public function get sold off to the highest bidder. Right now, the highest bidder for social media seems to be political campaigners and eco-warriors. Don't be surprised when someday the highest bidders include the world's major petroleum businesses, who will finally figure out the game. On their side will be the fact that the world still relies heavily on fossil fuels. When you see big oil throwing in a free Alexa (Oilexa?) with a fill-up, run for the hills.

13.
Lying with statistics

We're all used to having smoke blown up our asses by politicians and special interest groups. We develop early on a sense of when we're being used in this way, although some learn much faster than others do. Even from the inside—that is, when listening to someone we wholeheartedly agree with—we can or should be able to sift through what we're hearing and sort out what's true and what's just opinionated BS. It may make us feel good to have a conversation with someone who is absolutely uncritical of our ideas, but we should at some point be aware that we may just be succumbing to the reinforcing influence of surrounding ourselves with agreeable people. At the very least, we should try to be cognizant of other views, because if we can't we just become part of the problem when we speak.

Just as we learn how to sort messages we hear into reasonable/not, we also unfortunately learn about how the world really works. Sometimes if we're lucky, we get an excellent teacher, even if from an unlikely source.

FROM THE HORSE'S MOUTH

I had an excellent teacher once who helped me understand the ways of the world. This chap was a university professor who taught a class that greatly irritated me. In the undergrad world, there seems to be an inevitability of sitting through courses that one has no interest in, or with a professor that has belief systems that run counter to our own. Most of us know that feeling, and for many that experience crystallizes the sometimes hardheaded and intractable political or religious views we'll carry for the rest of our days, either by exposure to those who agree with us wholeheartedly and enforce our beliefs, or we hear someone so far on the other end of the spectrum that we know they're idiots and we're right and always will be.

Sometimes though that theory is soundly put to the test when we encounter a thoughtful teacher who jars us and causes us to rethink something we've believed to be profoundly true. The contents of those lessons can be like an irritant, like a thistle in a sock, that will bug us for the rest of our days. Those experiences are actually good things, and we should count ourselves lucky to have them, because that irritation is actually a reminder that we may be clinging to a belief that we've found out to be either untrue or not as fantastic as we thought.

A personal life-long irritant for me grew out of a tax theory class, which was mandatory in the program. I do not like taxes, I do not like discussing taxes, I do not like understanding the theory of various taxes, and furthermore, the combination of the words "tax" and "theory" makes me spontaneously combust. It has been to my stark horror then to realize that the class provided a truly effective understanding of a basic phenomenon that underpins much of the media-driven climate-war madness we see today.

It pains me still to take myself back to that class; admit it, the thought of listening to tax theory for a full semester must give you dry heaves as well. Those miserable hours sapped any pleasantness from my soul, and learning becomes even more challenging when sitting in a puddle of rage. I can still see the prof's weasel-like face, a scruffy white beard masking a countenance that epitomized shiftiness, deviousness, and cunning. His steady head held eyes that darted around the room like a lizard's, or as if in constant state of spatial analysis of the location of all exits, a trait I was sure he'd needed in his miserable line of work. Reinforcing my theories was the knowledge, learned early on, that the man was not an academic at all but had been a policy advisor for a number of governments. What a way to make a living, like repossessing dialysis machines from laid off workers. My initial classes were spent in a futile attempt to remain academically focused, which was impossible as I could not hide my loathing. Unfortunately, my searing gaze from third row far left never quite succeeded in vaporizing him.

Then the bastard totally upset my bile factory by turning out to be thoughtful and useful. Not just marginally so, like providing knowledge of how an engine works, but remarkably so. His experience in the world of politics, political reasoning, and purposeful communication provided incredible insight into how things work and why. He put aside in coldly logical fashion whether these ways are good or not, and focused on the way they are. I still hate his guts to this day, I'm not a big person, but I grudgingly tip my hat to him for the wise life lessons and admit that he probably isn't a total dead weight on society.

The most memorable of his examples was simple enough on the surface. He showed footage of a famous politician standing up in parliament, excoriating the prime minister of the day for letting corporations evade their fiscal responsibilities by refusing to pay all the "deferred taxes" on their corporate balance sheets. He gave a few large examples, noting what a difference it would make to Canada's finances if these amounts were collected, and asked why the government was giving a free ride to big business.

The politician asking the hard questions was Ed Broadbent, whose name doesn't matter at all I suppose but you can look him up if you want. He has a PhD in political science and was even a fellow at the University of Oxford for a year, which sounds silly to me but I suppose it wouldn't if I was one too. At any rate, he was a learned man.

Our tax prof then walked us through what was significant about the politician's demand. "Deferred taxes" are not real in any sense of the word; they are a figment of the imagination of accountants in order to justify their livelihood (they tend to dispute that conclusion, for some reason). When one calculates taxes in the accounting world, the rules are different from when one calculates taxes that need to be paid to the government. Governments allow taxpayers to write things off at different rates or in different ways than the accounting gods think you should. It's all complicated and unhelpful, and to get further into it will do no good. Suffice it to say that deferred taxes have nothing to do with what a company pays or should pay in taxes. They are simply irrelevant as far as financing the country goes.

So, our prof noted, Mr. Broadbent is clearly not an imbecile. Why then would he stand up in parliament and demand something as ludicrous as requesting that the government collect fictitious taxes?

This, the prof said, was the heart of politicized discourse. The point of the exercise was to make the prime minister look bad and to get a headline. The truth was not important. A startling (to me anyway) sidebar was the implied calculation that Mr. Broadbent and his handlers had made: that it was worth the risk of having some accountant somewhere stand up and say "that question was gibberish" in order to embarrass the prime minister. The gamble was shrewd and correct, because no one in parliament was well enough versed in the arcane world of accounting to point out the trick.

Before leaving that foul classroom, another subtle but critical insight was proffered in the class. The prof showed us the rules of a proper debate. Or rather, he had us try it first then explained the proper way to do it after. My tactic was to concede one of my opponent's lesser points, lull him/her into a false sense of security, and then crush them with an intellectual haymaker on the main point. I lost the debate, according to the prof, because I never recovered from the initial fatal misstep. The professor happily pointed it out as a textbook example of what you never do: you never admit that anything your opponent says is right. Ever.

Putting those two key lessons together, we can begin to see windows into why the energy wars of today are so hopeless. On one hand, we have a monolithic industry that for decades was welcomed by any government in the world and answered really to no one except shareholders. The world needed petroleum, countries needed petroleum, but also needed their resources developed. Oil companies were the only ones that could do it, the only ones with the know-how.

On the other hand, the protest/environmental movement had adopted the tricks of politicians. It works quite well to lob factually incorrect hand grenades into the theatre of public opinion, as long as those actually in the know are not nearly nimble enough to catch it in time. Subsequently, when they do, Part II of the strategy kicks in: never, ever admit any of the previous posturing was wrong. Simply move on to the next topic.

If that sounds implausible to you, let's look at a gloriously huge example from the recent past.

TACTICAL GENIUS – THE PLANTING OF THE 'OIL SANDS CARBON BOMB' MYTHICAL TREE

Everyone out there is familiar with Canada's oil sands, right? I know you are, because the whole world is. Go ahead, ask a herdsperson on the Mongolian-Manchurian steppes and they'll give you an earful about how bad the Canadian oil sands are for the climate. There is the possibility that they won't have heard of the oil sands, but say "tar sands" and the light bulb will go on as sure as the sun will come up tomorrow, because the pejorative and inaccurate label has been quietly but permanently inserted into the mainstream media.

How did this state of affairs come about? The oil sands are indeed a huge petroleum deposit, but so is the Orinoco Belt. Ever heard of that? What country is that in? Some will know it is in Venezuela and is one of the world's largest. Well then, how about Ghawar? How about Daqing? Those are the biggest fields in Saudi Arabia and China, respectively. Ghawar in particular should be memorable because it is the lynchpin of Saudi oil policy, their bedrock of production, and has been for more than fifty years. Yet hardly anyone knows about it specifically, or any of the other huge fields that have kept the show going for decades.

Given what the tax professor showed us, it became easier to see how the oil sands became the international pariah they are. The opponents of fossil fuels pulled a Broadbent. The oil sands' satanic reputation resulted from a few well-placed "deferred tax" type questions or hypotheses mused about in public arenas by just the right people, and then the right people's right people (the ones who were standing by to launch massive social media carpet-bomb runs of their utterances). From there, the oil sands didn't have a chance.

If you want to build up a myth so that it becomes conventional wisdom, you would first need the right people to propagate it. For the oil sands story, one typical and powerful example was generated by two of the most venerated names in the world of science: the publication Scientific American, and a NASA scientist. If a proclamation comes out that is supported by both the acclaimed voice of science and by an institution that's put people on the moon, it will have more credibility than if it came from the Pope himself.

In a noted 2013 Scientific American article[1] that quoted the NASA scientist James Hansen, as well as a few other science celebrities, the oil sands were found guilty in a court of scientific opinion, announced with the following verdict: "Alberta's oil sands represent a significant tonnage of carbon. With today's technology, there are roughly 170 billion barrels of oil to be recovered in the tar sands, and an additional 1.63 trillion barrels worth underground if every last bit of bitumen could be separated from sand. 'The amount of CO2 locked up in Alberta tar sands is enormous,' notes mechanical engineer John Abraham of the University of Saint Thomas in Minnesota, another signer of the Keystone protest letter from scientists. 'If we burn all the tar sand oil, the temperature rise, just from burning that tar sand, will be half of what we've already seen'—an estimated additional nearly 0.4 degree C from Alberta alone."

With that magisterial proclamation, these key messages hit the social media world like a proverbial plague of locusts, and the fate of the oil sands was sealed. Henceforth, people around the world would feel sick at the mention of the phrase "oil sands" or even worse at the incorrect but more efficiently damning "tar sands." People from multiple nations would be willing to break the law or chain themselves to equipment and blockade pipeline construction to prevent oil sands oil from ever getting to market. The European Union considered banning it. To this day, the connotation is so bad that googling "stop tar sands oil" brings up nearly 7 million hits, with websites devoted exclusively to doing just that. A certain percentage of the population has poll-communicated that they would be willing to be arrested to prevent it.

What makes this a Broadbent moment though is that the entire story is as fundamentally untrue as the effort to collect imaginary taxes.

It doesn't take much analysis to see this, and it isn't being particularly partisan to do so. The statistic mentioned in the Scientific American article about the global temperature increase was the catalyst to spark this revolution. To the layperson, this does indeed sound damning beyond words. But it is fiction of the highest order.

The oil sands quantity Scientific American refers to "if we burn all the tar sands oil" is 1.8 trillion barrels. This amount does indeed hold enough carbon to probably raise the world's temperature with enough left over to make enough diamonds to marry off every maiden in the USA. But there are no feasible, logical, or possible ways to burn all the oil there. It would be a feat of monumental proportions to produce even ten percent of that total.

We can see this easily, without fear of hyperbole and with no intervention from the fossil fuel industry whatsoever. Total oil sands production in 2017 was about 2.5 million barrels per day, and this was achieved over a decade of frenzied activity and well over a hundred billion dollars. There is currently no plausible scenario that would see oil sands production rising above 4 million barrels per day, and to get to that level would require vast new infrastructure simply to get it out of the region. That level would also cause the industry to exceed a total oil sands emissions cap that the provincial government put in place. So, 4 million barrels per day is beyond a realistic maximum, but let's use it as a devil's advocate number.

At 4 million barrels per day, it would take 1,300 years to produce all the oil in the oil sands. The area therefore, even at a production level 60 percent above today's, would have no material impact on global CO_2 emissions.

Oh wait, there is another arrow in the quiver: later in the article, when the authors presumably realize the above message is possibly shaky, they point out that another major reason oil sands oil is such a problem is because it is so energy intensive to extract: "Producing and processing tar sands oil results in roughly 14 percent more greenhouse gas emissions than the average oil used in the U.S."

Okay, at least that's real. Heavy oil does require more energy to produce and process than the average oil used in the U.S. But, what does that 14 percent mean, and what oil is below average, and how does that 14 percent number stack up against, say, heavy oil produced in Venezuela or Mexico or Iran or California? Is Middle East oil 14 percent less energy intensive if one counts the military footprint required to keep the whole region from dissolving into one big battlefield? You will not find any analysis of these questions, though you would think it would be completely relevant when singling out an oil deposit as a poster child for productive inefficiency and imminent global demise.

Finally, how do we know that *they* know they are being deceptive in using the Broadbent shuffle; that is, how can we be sure they are well aware of the false advertising and not just mistaken? Because it says so in the article itself. At one point, the author acknowledges that the 1.8 trillion-barrel number is not likely to ever be extracted, saying "even if just the oil sands recoverable with today's technology get burned, 22 billion metric tons of carbon would reach the sky." This number is less than 10 percent of the total ascribed to the whole 1.8T barrels. How does this prove that the intent of the piece is to deceive with known inaccuracies? Because the article leads off with the headline numbers—what would happen if 1.8 T barrels are dug up and burned—and that is the key message that is allowed to float out to the world. Not just allowed, encouraged. They tied the big, impossible number to the "carbon bomb" appellation, and sent that off to scare the world, rather than the more realistic number. If the 22 billion metric ton number is intended to earn the oil sands the "carbon bomb" appellation, consider that at 4 million barrels per day, it would take 116 years to produce 170 billion barrels to generate that emissions total.

The architects of this plan knew that a sensationalistic headline number would achieve the objective—that the oil sands are a "carbon bomb that will destroy the climate if it goes off"—and that any attempt to challenge this fallacy will be so paltry and late that it will be absolutely inconsequential.

After this, the second part of the wise professor's advice from the debating lesson kicks in: never ever concede an opponent's point. This is a particularly effective tool, because there is no room in social media for dissecting logic, so the insinuation just hangs there, planting seeds of doubt at no cost to the accuser.

As was also discussed earlier, the oil industry is no match for this caliber of chess. Some might call that fitting payback for the many decades when big oil called the shots around the world, and that these victories by environmentalists are simply the little guy finally winning against the big guys. That may be so, in a certain sense, but it doesn't make it right, nor does it make it any more palatable to see underhandedness win so easily.

WHAT ABOUT THE OTHER SIDE?

The preceding example doesn't mean that the petroleum industry hasn't used statistics to advance agendas either, it's just that the arena is or has been different.

The circumstances around petroleum's usage of statistics has been more direct and spoken to politicians' hearts, and indirectly to the population as a whole. Energy development means economic growth, lots of high paying jobs, lots of tax revenue, and an entrepreneurial industry. Governments tend to really like all these things. As such, the marketing job of the petroleum industry was massively different from the emotional pull of those who oppose it. Petroleum companies could bring a whole basket of goodies to the party, and governments seldom need much convincing.

There is a problem with that structure, though. The petroleum industry tended not to care a whole lot about what the general population was thinking about, because it didn't really matter. The consumers of the world needed fuel, lots of it, and in ever-growing amounts. Yes, there was always complaining when the price of gasoline started increasing, but the solution to that was obvious: people of the world, stay off our backs so we can find more of it. You hurt me, and it will hurt you more. This isn't a feedback loop, because the public became distanced from the very industry that provided the cheap fuel that underpinned the entire modern way of life. If anything, the distance simply turned to animosity, because the essence of the whole petroleum business was boiled down to one screaming, large, and ever-present message: the huge three-digit signs at every gas station that announce the price of gasoline.

It is lamentable that the sum knowledge of what the consuming public knows about its most important energy source is confined to those big shouting billboards with prices down to the nearest cent. As we've seen though, that corporate disdain for better disclosure led to a big problem. While the world focused its animosity on the petroleum business for gouging them, that was in one sense okay, because what choice did they have? Hate away. With the rise of the climate change movement though, including widespread civilian and governmental support, the times have changed. The petroleum industry now finds itself on its heels with a world that doesn't understand its business, and with the wolves at the door, it is a particularly challenging time to try to explain it.

HALF-TRUTHS HAVE MANY FORMS

We've looked at the current art of key messaging and the ways that various groups take advantage of the public's lack of understanding or attention to scare them witless. That angle works for a while, but after a certain number of campaigns, it loses effectiveness, like the old parable about the boy who cried wolf.

The same effect happens with the appalling output of the professional public relations industry. Their singular focus on saccharine key messages begins to lose impact, if it ever had any. A finely-honed key message that hits all the right notes and gets the thumbs up from the legal department is safe but at best useless and at worst annoying.

Take for example the statement made by Enbridge in their annual report concerning the Michigan oil spill that fouled a creek from one of their unhealthy pipelines. We'll look at the spill in more detail in a subsequent chapter, but for now, to refresh your memory, the incident was an embarrassing fiasco in which pipeline operating staff, operating a rickety old pipeline with a catalogue of thousands of flaws, overrode warnings twice and kept pumping oil through the ruptured line. A 2005 report identified 15,000 defects, of which about 900 were dug up, and that the pipeline operators come across as, based on reports of how operators handled warnings,[2] a room full of potheads than a well-trained operating crew . This was all that the public needed to know about the spill. If you heard about a big pipeline spill, and that the line in question sounded as though it was as corroded as a washing machine pulled from a lake after twenty years, and that the crew repeatedly restarted a line by overriding all the flashing red lights, you would be fairly certain what caused the problem. You would expect to hear from the company, at a minimum, something like, "well, that was pretty embarrassing and we are sorry beyond belief." But no, the company gave us a textbook display of state of the art public relations excrement. This is from their annual report that covered the year of the spill: "We strive to be a good neighbour in all the communities where we operate. In July 2010, when we experienced the most serious environmental incident in our long history—the leak of approximately 20,000 barrels of crude oil on our Line 6B pipeline near Marshall, Michigan—we immediately accepted responsibility and made a clear commitment to local residents to clean up the spill and address the impacts on the environment and to the people of Marshall, Battle Creek and area. We are determined at all times to meet our responsibilities, and live up to and exceed the expectations of our stakeholders." Good lord. An eight-year-old would know that when you're caught with your hand in the cookie jar, it's not wise to say that stealing cookies is the worst crime you can think of and that your number one priority is to do what your mother says, and that your goal is to exceed her expectations. Yet there it was in black and white, a statement so poignantly ineffective it almost brought tears, like watching a simpleton try to coax a dead bird back to life. To be fair, Enbridge did accept responsibility, but for the spill's victims, the statement sounds more like humility-free bragging than any sort of apology.

One might say that is the price to be paid for being a public company in the litigious world we live in, that Enbridge said the right things to safeguard the interests of shareholders, or whatever. That's one view, and it is valid if your goal in life is not to be sued.

Here is the way a class act does it, with honesty and integrity, and this person is equally as encumbered by capital markets. Here is the legendary Warren Buffett describing mistakes he has made. See if you can spot any differences between his mea culpa and the one Enbridge crafted. This segment is from Berkshire Hathaway's annual report[3]: "The crowd of companies in this section sells products ranging from lollipops to jet airplanes. Some of these businesses, measured by earnings on unleveraged net tangible assets, enjoy terrific economics, producing profits that run from 25 percent after-tax to far more than 100 percent. Others generate good returns in the area of 12 percent to 20 percent. A few, however, have very poor returns, a result of some serious mistakes I made in my job of capital allocation. I was not misled: I simply was wrong in my evaluation of the economic dynamics of the company or the industry in which it operated... Fortunately, my blunders usually involved relatively small acquisitions. Our large buys have generally worked out well and, in a few cases, more than well. I have not, however, made my last mistake in purchasing either businesses or stocks. Not everything works out as planned."

Which one would you buy a used car from? Which would you want as a neighbor? Which has more credibility when something goes wrong? Both accept responsibility for errors, but one party clearly doesn't get it, and should be in no doubt whatsoever as to why public trust has eroded in them and whoever else follows the textbook.

A long time ago, I heard a valuable bit of advice to live by: "A person should be upright, not kept upright." As a first step in bridging the gap in the fossil fuel wars, all involved should get that tattooed on the inside of their eyelids.

14.
Not lying with statistics

The fossil fuel industry has gyrated itself into a bit of a predicament. After decades of doing pretty much whatever it wanted on a global scale, it now is in a position where it has to explain itself and defend not just its practices, but its existence.

The success of the fossil fuel business has, in a way, compounded the problem of getting the message across. Petroleum products are so entrenched and ubiquitous that it defies credulity to try to explain to people what would be left of their comfortable worlds without fossil fuels. The list of consumer products is almost all-encompassing, and the things that aren't on the list were likely brought to you via fossil fuels.

We then get to a point where if the petroleum industry gets on a soapbox and says, "You know, without fossil fuels you wouldn't even have lipstick," the reaction is pretty much a cow-like returned gaze. The general population, the segment that wears lipstick anyway, most likely would not in a dozen years spend time thinking about the connection, and for the few that do, there is nothing to be gained in pursuing the line of thought any further. So what? Is that a threat to withhold fossil fuels from lipstick production? What are the odds of that? If it registers at all, the connection is at best a "well I'll be damned" sort of reaction, and the factoid is filed in the mind's Great Drawer of Useless Facts, never to see the light of day again.

Compounding this problem is another foible of a large slice of the population: the fact that numbers and statistics are utterly fascinating to a few, but an annoying black box to many. Not only is it hard to explain how dependent we are on fossil fuels; to convey the scale of the issue presents a particularly large challenge. The challenge is so big, in fact, and the ability to grasp these abstracts so rare, that some companies have seized on the ability to accurately assess these macro variables as a key hiring characteristic.

TESTING, TESTING

Tech companies have not been around for long in the big scheme of things but have already undergone a number of cultural revolutions. Most won't remember, but from the 1950s to the 1970s, technology behemoths like IBM were staffed by squadrons of square-haircutted men wearing white shirts and suits and ties. Not colored shirts, white shirts. They looked professional and clever and undertaker-ish, and they dealt with global corporations and sold computers the size of basketball courts, and they all had the same white-bread look and came from the same business and tech schools. Unconformity was not just rare but inconceivable; one didn't show up for work with a blue shirt or messy hair. It just didn't happen. As modern technology took hold, however, it was clear that the end of that uniformity was in sight.

The industry then flipped over completely, and became famous for sandals-wearing employees who flopped around on couches and played ping pong to relieve the stress of having a job. Creativity was in, old-school thinking was out, and the industry became the place to be; nowhere else was the path to a billion dollars so direct.

An unusual habit arose in the hiring departments of these institutions. Whereas job candidates had historically been asked questions about skills, experience, dealing with jerks, and times where you did something amazing, tech companies began asking off the wall questions to try and get a better look at people's thought processes. Many of the questions are just plain weird, like, "Do you believe in bigfoot?" (which is both weird and dumb. Who doesn't?), while many were challenging numerically. How many square feet of pizza are eaten in the US every day? How many snow shovels are sold worldwide every year?

I have no idea whether this line of questioning produces good results or not; based on the frequency of crashing computers, I'd say no, but I could be wrong. What is interesting about this process is that it illuminates an issue that often puzzles bystanders, particularly in the energy world.

These questions about commonalities of our world, and the ability to extrapolate them, are fiendishly hard for the general population to master. It's not a matter of stupidity or ignorance or lazy thinking; it is simply a fact of the human condition that many people are susceptible to underweighting or overweighting various numerical realities. A significant portion of the population is stressed by having to calculate a restaurant tip in front of others, so it should be no surprise that global statistics, no matter how fascinating to some, are not very appealing to the majority. A book discussed earlier, Hans Rosling's excellent *Factfulness*, provides dozens of excellent examples of the general public's deficiency in accurately understanding the characteristics of the world they live in.

The impacts of this are quite significant. It is a problem because it is difficult to envision the actual scope and scale of abstract things, while at the same time it is easy to have our views manipulated. We therefore often misread some critical information. Two large examples rear their heads.

GLOBAL OIL PRODUCTION AND CONSUMPTION

As tech companies have been researching, people are not strong at making abstract global estimates of any sort. That applies to energy consumption numbers just as well as snow shovel sales, and possibly even more so. Energy consumption is so abstract that it is hard to quantify. How much electricity do we consume in a day or a month? We may not even know if we look at our power bill, because those things sure as hell aren't intuitive. What the heck is a "barrel of oil"? What does a hundred million barrels of oil look like? Is that the same as how much Coke the world drinks every day? Is it a hundred times more or a hundred times less?

What we will find when we talk to average people about oil production, for example, is that they've accumulated, if anything, some wildly dangerous numbers from the media. Concerning oil production, the predominant media narrative of today is all about the fracking revolution, and about shale oil, and that the US now dominates the global production scene. The theme is that the US is producing at record levels, which is true, and that new technology has unlocked resources that have turned the US from a major oil importer to a major oil exporter. The second order narrative, if anyone gets that far, is that OPEC is pretty much dead, and that the US can and will keep prices low because if OPEC tries to raise oil prices, the US will flood the market and lower them again.

Before discussing why this impression is so dangerous, it's worth flagging how wrong it is. The US has indeed undergone an oil production transformation, and has doubled output in the past 10-15 years. That is big news. But on the global scale? US oil production has gone from about five million barrels per day to just over ten, due to shale production. That's big news for the US, but the world produces and consumes about a hundred million barrels per day. Therefore, the US shale revolution has, at a cost of between half and one trillion dollars since its inception a decade ago, added 5-7 percent to global production. It will take everything the US has to add another 5 percent. Has the US actually become an oil exporter? That stat is another example of smoke and mirrors. It is true that the US exports oil, and only started several years ago, but that's because it was illegal to do so until then – the US government had made the practice illegal decades before, and only relaxed the law in the past decade as shale production grew. Furthermore, the amounts exported are far less than the amounts still imported. The US is therefore still a net importer of oil, and it is not self-reliant as is often advertised.

The story is an example of the ethnocentric nature of thinking that what happens in our line of vision is what's happening in the world. It is dangerous when individuals succumb to this, it is potentially lethal when countries do.

What makes this so harmful? When the media accepts these narratives as true, everyone believes it and starts acting that way. Individually, we see consumers buying big trucks and SUVs because the price of oil is currently not all that high. Various industries see natural gas as the lowest cost fuel around (which it is, for now), and start switching en masse to the stuff. Governments start flexing their muscles and acting as though they don't need any other nations (not pointing at all at any funny-haired and belligerent world leaders, no, not at all). As these notions become entrenched though, what happens if they're not right?

That last point is important, because they're not. The US is not a net oil exporter, nor is it self-reliant, and the belief that shale oil will make it so is untrue. The US is drilling up its shale sweet spots at an aggressive pace, and the party won't last forever. Furthermore, with only a fraction of global production, the US is an important producer but not as dominant as believed. Any policies built on the notion that they are dominant may well unravel in an unfortunate manner, a side effect of not understanding the key global statistics—not just the ones we can see by looking out the window.

RENEWABLE ENERGY PRODUCTION AND CONSUMPTION

If petroleum statistics are hard to fathom, renewable energy ones are utterly impossible. As with petroleum misinterpretations, there are potentially serious consequences for misreading the significance of renewable energy in the global mix.

The most common media narrative with respect to energy, the dominant theme of all the big information flows, is how green energy is progressing. Sorry, not progressing, exploding. It is common to read stories about the vast quantities of renewable energy capacity being added, or the percentage that some country gets from renewable energy (which is usually astonishingly large).

The problem is, again, that the narrative becomes stuck in people's heads, supported by dubious statistics, and we never understand the true meaning of them. Take for example the statistic about how much energy some countries get from solar power. In Germany, the number can be quite large. One headline from January 5, 2018, announced that renewables had covered 100 percent of Germany's power usage for the first time *ever*. The statistic was true, but under exceptional circumstances, and for only a flash of time. The achievement took place at 6 am on January 1, after the nation had passed out from New Year's Eve celebrations and power usage at an extreme low. It takes little energy to get up and vomit and go back to bed. At the same time, the wind happened to be blowing exceptionally hard. It is generally not acceptable to cherry-pick confluences like that; if it was common, all sorts of bizarre records could be set. At any point, there is a confluence of some set of factors that make the point unique. It simply isn't newsworthy to flag these infinite coincident peaks—unless the topic is renewable energy, apparently.

The message that sticks in the minds of the public then is that Germany can get all its power supplied by renewables, with the corollary message that fossil fuels are not necessary. Nothing could be further from the truth. Under almost all circumstances, fossil fuels are not just required, they dominate German electrical production. That's because renewable usage tends to dominate when demand is nowhere near the peak, which makes for nifty headlines, but the country still needs a full arsenal of fossil fuel-powered plants to meet electrical needs.

The public then becomes mentally anchored and acclimatized to the idea that we don't really need fossil fuels anymore, or that that state of affairs is imminent.

Perhaps you're thinking, so what? Why is this a problem? There are a great many things the public doesn't understand, from how to keep a public restroom from being revolting to how their car works. Or how their electricity gets there. Why should we think it relevant whether the public understands any of this?

WE GET IN BIG TROUBLE BY NOT UNDERSTANDING THESE CONCEPTS

The answer to why we should care if the public understands has to do with the current abysmal state of discourse with respect to fossil fuels.

If we accept these flawed conclusions, such as that we live in an environment of perpetually cheap energy due to the shale revolution, or that renewable energy has come so far that nations can now exist solely on it, we are fooling ourselves into acting in ways that are not conducive to our survival. That may sound melodramatic, but it is true.

If we believe that the world's fossil fuel needs are satisfied for many decades to come by the US' petroleum production capability, and that the fuel will remain cheap because of these technological advances, we tend not to value efficiency. We tend to relax and consume indiscriminately. That's simply how humans work. When the cost of something is low to the point of being immaterial, we consume it without even noticing. Consider plastic drinking straws, made indirectly from hydrocarbons. Have you ever thought about them? Have you ever really valued them, or thought maybe I should keep these handy things and use them again? No, we are handed them for free and we treat them as such. We've actually come to expect them to be wrapped in their own little sanitary plastic or paper wrapping, which we discard even quicker than the straw.

Of course, a straw doesn't matter to the environment, but a billion of them do, and we find ourselves right back at the problem that people either don't or can't often think of the magnitude of these large numbers. Imagine if all the drinking straws thrown in the waste in a single day all over the globe were dumped in your living room. Does that help with the context? Because it becomes apparent without too much imagination that a day's worth would fill a gymnasium.

And that's just drinking straws, which we abuse because they're free. What does it mean when we treat other things as near-free? What about cheap gasoline, and what that does to our vehicle choices? We opt for bigger vehicles with more creature comforts, which both burn more fuel and require more to manufacture. We fly more often, adding voluminously to the global emissions tally.

What becomes significant for the globe is that millions and millions of others think this way too, with extremely serious consequences for our consumption and waste patterns.

What about renewable energy then? What's the harm in believing that we are either at or near a point where renewables meet all our energy needs? That one is equally dangerous, but is a bit more complex.

AGAIN WITH THE SYSTEMS

If we are exposed to enough of these headlines that plant the seed that we are on the verge of existing solely on renewable energy, we anchor our expectations to that concept, which is another universal human trait that subconsciously appears. When we become anchored to that idea, we then believe that we really don't need fossil fuels anymore, and the narratives put forward by the anti-fossil fuel movement seem logical and appropriate. They even seem like excellent ideas because, coincidentally, we can clearly live without fossil fuels and if we eradicate them, we will save the climate! OMG, look how wonderfully that works and it wasn't that hard at all.

But you can feel it coming, right? The impending arrival of the bucket of cold water. Is it really that easy for a country like Germany to be free of fossil fuels? Have they actually done that, or have we been hornswoggled?

Hornswoggled might be a bit strong, but it is a pretty fun word so it can stay. The media headlines stating that Germany now gets 100 percent of its energy needs from renewables isn't meant necessarily to deceive you, it is simply what the media does to attract eyeballs. A bit of harmless arm-waving in the eyes of editors whose job it is to attract viewers.

However, the cumulative impression it leaves is far more significant, and we know the problems with actually accomplishing what those headlines imply. There is no way that Germany's, or any other country's systems can adapt that quickly to convert entirely to renewables. It's not even close. The entire country could be blanketed with solar panels, and wind turbines that are made of solar panels, and the country would still be reliant on fossil fuels for more than half of their total energy consumption. And that's speaking only of electrical requirements. That doesn't at all account for heating needs, which are almost universally met by burning fossil fuels.

There is a price to be paid for all of these misinformed views. We are convinced that some things are not to be worried about at all, such as whether we have enough energy or resources to meet our growing needs. We are force-fed messages that make us fearful about, namely, fossil fuels.

As it turns out, the fear narrative also works well in the social media world, because humans are also incredibly bad at something else: understanding what really is risky, and what just appears so.

15.
Fears are real, but don't always make sense

Someday we're all going to get that tap on the shoulder from the grim reaper, who will say "You're coming with me, buddy." If we're lucky, it will be when we're so old that we greet him or her with, "Where the hell have you been? One more diaper and I was going to find you myself." Unfortunately, some will lose their life in battles with diseases or tragic accidents. Others will take a different path, exiting in a YouTube induced blaze of miscalculated glory that illustrates the wrong-headed ingenuity of modern thrill-seekers that we try hard not to laugh at.

The point is, we all go at some time and we have no idea when that will be, but we are fiercely addicted to living and often go to great lengths to remain that way. In other words, we are often militantly aware of perceived risk. This awareness is ingrained from our ancestors' days as cave dwellers and is as near universal as almost any human trait. Fear has kept us around for a long time, longer than perhaps we deserved given our propensity to war, showboat, and overeat.

Our fears can also be spectacularly wrong. Indeed, a great many things can harm or kill us, and some of them we can spot rather well, but for many, our brain plays tricks on us, inappropriately overweighing some while doing the opposite for others.

Fear is an ugly topic to tackle in this book because it could fill volumes. We will have to narrow the scope substantially, not even getting to the fun ones like a fear of balloons. As much as that is clearly a cry for some kind of help, we must turn our backs on that unfortunate few in the cruelest possible way. We have to. Fossil fuels are at the heart of many fears on their own, providing plenty to think about without trying to resolve fears of childhood staples, which are likely the bread and butter for many a psychotherapist, and maybe even the highlight of their day. We'll give them that and stick with fossil fuels.

Before getting to fossil fuels, though, it is unfortunately necessary to wade into the realm of the more common human fears, because there is a direct path from that mess to the fossil fuel wars.

GENERAL LIFE FEAR

Outside of the oddball cases, human fears are shaped essentially by two main factors: the basic instinctual ones, and the ones that we allow to grow and fester in our minds. I am sure that any psychologist within earshot would be howling with indignation at that superficial description, but there is no need to muddy these waters with diminishing-returns hair-splitting.

Instinctually, we are scared of the dark, scared of animals with big pointy teeth, and scared of lightning. None of these are necessarily irrational; in fact, they are particularly helpful. We, as a largely urbanized society, may have strayed from random attacks by wild beasts, but then again, I suppose it depends on what you label a wild beast. How do you classify Harvey Weinstein, for example? Suffice it to say that fears that harken back to our basic survival days are handy items in the big tool kit of life.

Now, let's turn to the flip side, the fears that are engendered not just by modern life, but modern information streams, specifically entertainment streams. Between the two, our fear map has been rewritten and it is fair to say that we abandon a lot of rationality in the process, and allow the part of our psyche that is impressed by sensationalism to override the more rational part. Fear as entertainment and sales tool has become a part of modern culture. The purveyors of fear are expert at pushing our panic buttons, and usually with the express purpose of emptying our wallet in one way or another.

Fear as entertainment shows up a lot on TV and in movies, where we tune in, for example, to endless crime-dramas. If the gruesome murder techniques don't scare us outright, the sheer frequency should. There are a hundred serial killers caught every week on TV, a considerably higher

FEARS ARE REAL, BUT DON'T ALWAYS MAKE SENSE

number than there seems to be in real life. No other theme is so consistently rammed down our throats, and we don't seem to mind one bit.

We are also sold fear for other purposes. It is used as a political motivator, where politicians appeal to our baser instincts in order to ensure voters that they are the choice that will offer safety and comfort. We are also sold fear to make us buy products.

We can see the effects of all the scaring show up in weird places. Consider an example from the endless greyish-beige of the upscale shantytowns that make up suburbia. Let's enter a typical one, and let's call it Gorky Park, because it is more realistic than the kind of stupid name developers attach to these ill-conceived hyper-lazy visions of middle class paradise such as Tuscany or Hamptons or Somerset, without any sense of irony whatsoever (spoken as a resident of one).

At any rate, the streets of these ghastly subdivisions have at least one benefit, which isn't immediately apparent: they are nearly impossible to navigate due to the faux-character imbued by some dull-witted CAD operator thirty years ago. A geometrical progression of loops and arcs and cul-de-sacs feed into one main artery of ingress/egress. That may not sound like a benefit, but it is in one sense: criminals are almost universally uninterested in casing Gorky Park or any of the other thousand facsimiles. Who would want to enter a maze like that, find an unoccupied house (not easy to spot, because they all look equally lifeless), rummage through it, find the few valuables inevitably hidden under socks or underwear or in a box in the closet, flee the house, and then promptly become lost and disoriented while trying to exit the neighborhood?

This speculation is borne out by the fact that in twenty years in the neighbourhood in which I reside, no crime of significance has occurred. Most other suburbs are similar, according to the crime hot-spot maps the city puts out. A purse and an old ratty bike are the only objects I can recall going missing in two decades; the purse was found on the street and the bike was found a dozen houses away at the top of a set of stairs, the master criminals having been foiled by this insurmountable obstacle.

What does all this have to do with fear? Most inhabitants of such neighborhoods rush into their homes and lock the doors behind them, day or night. This is borne of a fear derived from watching a lot of TV shows and movies about random killings and kidnappings. Based on the number of these shows and how often they air, one would be forgiven for thinking these crimes happen on every street every other week. But they virtually never happen. Most crimes that do happen are related to the drug trade or other criminal activity; the odds of a stranger randomly attacking or invading a suburban house are as close to zero as they can possibly be. Yet we live in our beige boxes as though they are bunkers, fretting about the efficacy of security systems that the police won't even bother responding to, and swiftly dead-bolting doors at two in the afternoon as if in the middle of an apocalyptic invasion.

This is but one example of how we let media portrayals of danger rule our lives. We hear about a horrible crime and our imagination pictures it happening to us, and we take all precautions we can to ensure it doesn't happen. In a city of a million people, one weird random act of violence is just that, a one in a million event. Living in the equivalent of a cornfield maze only makes the odds decrease.

There are countless other examples. They are simple to test for: if you feel ill at ease about something, consider whether it's been the subject of a TV show. If it has, your concern is most likely out of proportion to the actual chance of it occurring.

And what of the things we really should be worried about?

WHAT SHOULD SCARE US BUT DOESN'T

The other day I joined a new club, one that bestows on me a certain modern authenticity as a truly connected individual. It's not that hard to join, but it's not free. The price to join has a somewhat thrilling scalability because it is unknown beforehand. It is not necessarily a monetary cost; it is more psychological in nature. The club is fairly new and involves some physical challenges that give a jolt of adrenaline. The initiation experience may not be physically fun—in fact, it may cause physical discomfort or even harm, not unlike many adventure sports. That's the buzz of living on the edge, the part that wants to cliff dive to feel alive. We know there's danger and the price could be high, but hey, some of us live our lives in fear and solitude, and some of us push the envelope, just like in the commercials.

FEARS ARE REAL, BUT DON'T ALWAYS MAKE SENSE

What happened was, while walking down the street and reading a text, I walked into a tree. It was a minor hit, not as dramatic as a full-on head shot, but an errant small branch with surprising longitudinal stiffness found the side of my neck, leaving a bright red furrow a few inches in length. I am now one of *those people*.

What is the cost of joining this club? It depends entirely on how many people were watching. Was I going to be immortalized in a YouTube compilation sequence, or heaven forbid, become a meme? To us living-on-the-edge warriors, pain is nothing until we ascertain how many witnesses there are. Mercifully, the price I paid was low; few people were around, no one appeared to be filming, and no one even seemed to notice. Once that was clear, I could then stop pretending like nothing happened and whimper quietly as I investigated my wound. It was just a scratch, only a flesh wound. It missed my eye by a mile.

Before picking up that thread again, let's head for the web for a second to check out a different eye-safety angle. A quick search for "racquetball protective eyewear" brings up a stylish and assorted lineup of options from $24.99 and up. An interesting side note about, or characteristic of, racquetball, which is played on enclosed courts, is that the courts are universally white to aid with visibility, so that the ball is easier to see, and there is no other activity in the game other than watching the ball with all our concentration. However, to be on the safe side, an entire industry provides protective eyewear for racquet sports, because what kind of fool risks their eyesight for a dumb sport?

Welcome to the modern world of risk assessment, where we consider a flying rubber ball that we're watching intently more dangerous than sharp sticks at eye level we don't even bother to look for.

If I were some unique idiot, the only one that paid attention to a device while walking, that would be one thing. As can be determined within seconds though, it is an epidemic, and not just staring at our phones. We misunderstand risk in ways that are so flagrant and obvious that it's a wonder we reach adulthood. If I were a racquetball player, I no doubt would buy the protective eyewear, because that's what you do. Walking down the street and texting appears well within my bounds of reason, because that's what we do.

We safeguard against unlikely things where the potential danger is clearly evident and limited, and we blissfully ignore major risks that are life threatening. We go to our offices and apply hand sanitizer, then we cross the street so engaged in conversation that we walk right out into traffic without looking because a little light fixture across the street says it's okay to go. We text and drive right past people who are texting and walking. It's almost unbelievable to watch if one pays attention to it all, and I can say that as one who's done it and was fortunate not to lose an eye, and I am far from alone. I once had to slam on the brakes for a man and a woman jaywalking across the street right in front of my vehicle, a toddler between them and holding each parent's hand. I've seen more near-death experiences on the streets of downtown than in an earlier life of working with heavy equipment under dangerous and exhausting conditions, and I'm really not out on the streets that much.

This sort of blasé attitude towards certain elements of safety appears to arise because incidents are so infrequent—but so are murderous home invasions in the bowels of the suburbs. We clearly do a poor job of worrying about the right things.

FEAR OF CLIMATE CHANGE

When we combine the above factors—a tendency to allow ourselves to be scared by media coverage or semi-mythical boogeymen, and an inability to understand or act on what is truly risky in an appropriate way—does that leave us in a good position to understand the risks of what will happen to the world fifty years from now if global temperatures rise by several degrees?

Most likely, we will either succumb to the visions of apocalyptic terror that the media stream pushes, or we'll climb in our SUV and go for ice cream and not worry about a thing except the calories, and only worry about that after the feeding frenzy but never before.

Neither of these perspectives is helpful. Some would argue that the existence of civilization is at stake if we don't do something about climate change, and thus it is wise to instill fear in people. It is not hard to find horrific projections of what the world will look like if it warms by three or four degrees by the turn of the next century. Furthermore, anyone who ventures out on a limb to make these projections is not going to do it to say, meh, things will probably be about the same. The goal with these projections is to shake people out of their lethargic ways, to catalyze them to take emissions seriously.

What makes the fear-inducing tactics unhelpful is that they don't really work, although they work really well on ten percent of the population. The other ninety percent couldn't care less. The majority are like me walking down the street and texting. When that happens, it simply becomes annoying to be lectured if the threat is not imminent or even clear.

Perhaps the reason that the vast majority don't seem to care is that the problem is too abstract, and the task seems insurmountable. Average citizens know inconvenience when they see it, and it doesn't take much of a look around to see that what is being asked of them in order to prevent climate change is going to be very inconvenient indeed.

Give up fossil fuels? That might sound neat as a slogan, and yes, there are a lot of windmills and solar panels going up, but how will modern life work without good old petroleum and coal? There are countless examples: How would the grocery store get refilled without all those trucks, and how will the air conditioners run, and for that matter how will I heat my house? Do I have to give up vacations involving flights? Whoa, now let's get serious. Give up my absolutely necessary tropical holiday? I'll take the hotter planet, thanks.

That factor is the biggest barrier between the threat of climate change as posited by a few, and the realities of the life of the average person. The hatred of fossil fuels and demands that we get rid of them don't seem remotely realistic, because it is easy to think that, at this stage of the game, no one can. It's true, with our standard of living, no one can get off them completely, and more importantly, no one really wants to. If they do, they don't act like it. Environmentalists don't even seem to act like it. Try Googling "international climate conferences" and you'll see them in every exotic location on earth, and flying to them is the only possible choice.

The choice for consumers then becomes: feel terror and abject hypocrisy at the same time because fossil fuel consumption makes one a part of the problem; or say ah, to hell with it and let someone else do the worrying.

FEAR OF RUNNING OUT OF FOSSIL FUELS?

Forget FOMO. Here's something that should be just as worrisome as the fear of climate change: the fear of running out of fossil fuels. FOROOFF.

Running out of fossil fuels. That just sounds stupid, doesn't it? I've never heard of a single person having that fear.

That's not quite true. I do know a few people that, while not afraid of running out, are definitely aware of the consequences of running short, on a smaller scale. Have you ever worked with a safety specialist or emergency response planner? They have a unique view on the world. One that I am acquainted with always has a vehicle full of fuel; he fills up all his autos when the tank falls below half. He always has enough gas in one of his vehicles for his family to get to grandma's house five hours away. He knows that the tank of your toilet is an excellent source of potable water when all else runs out. He knows that most people don't have a backup water supply in case of a major emergency, and that in the event of one they'll be happy to access that source.

What a freak, hey? Or is he? What would you do if natural gas ran out in the dead of winter? Do you even know how your home is heated? What if there was some sort of mass emergency, as we see on the news sometimes, where basic staples of existence become unavailable? Think beyond gasoline: what about formula for your baby? Do you have a month's worth stocked up? What about diapers? Sounds crazy, right? But imagine our supply chains going still for even a few weeks.

We don't even think of those possibilities because they rarely happen in North America, but they easily could. The vulnerability of our infrastructure should make you go pale if you think about it. A few well-placed explosions on major pipelines could lead to absolute chaos, and those pipelines are startlingly unguarded and accessible to any terrorist with the ambition to visit a website, pick up a map, and drive around the countryside.

But those thoughts never enter our heads—ever—while we panic if we go outside without sunscreen or if a murder happens a mile from our home.

FOROOFF will never enter the lexicon, even though it should, because we have far more real fears like the fear of missing out. That's not an attempt at facetiousness, it's a fact. FOMO absolutely guides people's actions and is a completely real phenomenon. True, it's not a real fear like that of falling into a tank of sharks, but it is embarrassingly real as evidenced by the weighting we put on it.

At the end of the day, people will never be completely properly aligned with what they should be afraid of. There is no point in trying to change human nature, or the fact that effective marketing campaigns exploit this weakness successfully and always will. Earlier, I lamented the fact that focus groups totally misconstrued a key safety message related to pipelines, that it did not resonate with the group. Unfortunately, sometimes, there's no way around that either. We simply have to lay out facts as rationally as possible, and consistently and patiently work with the public to find ways to get the point across.

Concerning fossil fuels and fear, however, we can be certain of a few things: trying to live without them at present is something we absolutely should be fearful of, more so than most. That's not to say we shouldn't be concerned about an environmental future, but on this topic at least, it is imperative to understand what's at stake in the short term before worrying about potential and speculative scenarios five decades hence. A lot will change in the next half century, and guessing at those changes with an eye to extreme accuracy is the mark of a lunatic.

16.
Consumption and population are both growing

Have you ever been on a cruise ship? A great big porker, one that can haul three thousand people around for ten days? A gargantuan pleasure craft custom designed to make you gain weight, and offer you several sun-drenched acres devoted solely to showing the world just how much? That is actually one of the good things about cruise ships, in a sense: they are places where people can truly relax, more or less free of judgement, because almost everyone is the same. Just so you know, most of them are the same. I'm just telling you, most are the same. Brochures may indicate otherwise.

I have had the opportunity to go on a cruise, and waiting to board is a kind of a bovine experience. Now that I think about it, it is precisely a bovine experience, waiting in a long line that twists and turns, no escape allowed, all ready to be herded aboard a transportation vessel when our turn comes, just like those poor hooved beasts. The accommodations are definitely a step above what cows experience aboard their overland coaches, although the viewing portholes are eerily similar. One hopes the final destination of the human's journey is an improvement over what awaits the unfortunate cow.

While waiting in line for what seemed like hours but was actually only ninety minutes, I had a line of sight to the loading portal of the ship. This hole in the side of the boat was the only thing of interest within my view, other than the other thousands of people waiting in line that I'd be seeing for the next eight days. They had not yet begun to feed and relax so were uptight and unpleasant to look at. I focused on what the army of conspicuously foreign workers was doing in preparation for our lap of the Caribbean.

To maximize return on investment, these huge ships don't appear to hang around. The loading portal was a beehive, and through it went a staggering number of pallets of supplies. With nothing else to do, I tried to do a rough calculation of how much bacon alone was heading aboard, but I soon frightened myself and abandoned the effort. You will see what I mean if you consider 3,000 people in holiday mode, eating bacon five times as often as they normally do, and eating ten times as much as they normally do because it's free. I'll be honest with you: some of these were not small people.

There were waves of drinks and vegetables and general food, and boxes and boxes of consumer supplies. I can't even begin to imagine how much toilet paper. I believe I turned away at that point. I also didn't notice the alcohol stash, but surely it had to be huge as well.

What was interesting about this little side show was that it provided the ability to actually get a feel for how much a group of consumers can go through in one week. True, the consumption on a cruise ship is far more than average—lord help us all if it's not—but on the other hand, the consumption on the ship is largely limited to food and beverages and assorted paraphernalia. The cruise experience leaves behind ninety-nine percent of the stuff we consume that comes from the mall, or Home Depot, or Walmart. We are part of a food and product chain that is so large and energy-consuming that it is virtually unimaginable; it is amazing to see what a few thousand people go through in a week. To extrapolate that to seven billion people is virtually impossible, but we'll try anyway.

SOME THINGS WE CAN'T COMPREHEND DON'T MATTER, OTHERS MOST CERTAINLY DO

Before contemplating what a pile of the earth's daily food would look like, it is worthwhile again to turn to astrophysics. I know, I mocked puny human efforts to see the universe some chapters ago, and I'm not going back to the topic just to be cruel or to kick astrophysicists in their little moons. Outer space happens to offer a useful context by which to judge the enormity of certain things, and to make ourselves aware that some things right in front of us will always be incomprehensible.

Next time you're out camping somewhere with little light pollution, look up at the stars. To look at a star is actually an amazing experience. We are

CONSUMPTION AND POPULATION ARE BOTH GROWING

seeing, according to the big-telescope brigade, the light from an object so distant that we measure it in light years, the distance light travels in one year (about 6 trillion miles or 9.5 trillion kilometers). According to my twelve seconds of research, the nearest star (that is not the sun, don't look at that one) is about 4.4 light years away—over 25 trillion miles/40 trillion kilometers. Sit there for a second as you look at the little light, and try to make sense of that number. Recall how you sometimes hear a statistic that illustrates vast sums, such as that a stack of a billion dollar bills would be sixty-eight miles high. When we hear that, we get confused and disoriented by even trying to imagine Bill Gates' wallet. A light year, then, is something so far beyond our senses that it is almost meaningless.

Now let's look at distance another way to get to something we can relate to that can make a small distance seem huge. Say you're out camping and hear a rustle in the bushes. You cast an alarming glance at your tent or vehicle, which may be fifty feet away, and you almost crap your pants because that is not a large distance with a bear in hot pursuit.

At the end of the day, which of these distances meant anything to you? Despite our "bigger is better" way of living, the distance to the nearest star is utterly meaningless. We can't comprehend ten thousand miles, which would take us nearly halfway around the world, so a trillion miles loses all meaning. It's like counting the grains of sand on a beach. It's just another insanely large number.

Tech companies, which generally have the luxury of hiring the best and brightest, screen those people by asking them questions about the global consumption of a single item. That presents a challenge for the top-notch candidates they have the luxury of dealing with, so what hope do the rest of us have? Look in your garage or kitchen or living room or toiletries travel kit and imagine how many of each of these the world consumes every year. Try this for as many objects as you see lying around your home. Unless you are a hoarder of world-class proportions, the number of objects we're talking about in this simple thought experiment is but a tiny fraction of what we use in our lives.

What we need to be able to do, for the purposes of this discussion, is to contemplate what the systems look like that are required to provide all this stuff for the whole world. That part is critical if the discussion is to be about where we're at on the continuum of the use of fossil fuels. It does no good whatsoever to simply wave your arms around and tell people to stop using something when their whole life runs on it. With respect to our staples of life, it's a big deal if we don't understand it, because we not only take it for granted, we allow people to tell us they can bring them all to us in a totally different manner, one that has virtually no environmental footprint. People tell us they can replace all our fossil fuel requirements with wind turbines and solar panels. Who wouldn't sign up for that?

As long as fossil fuel consumption, or banana consumption for that matter, remains an abstract concept that we are can't easily picture and are therefore not interested in, we can be easily swayed by arguments that getting off fossil fuels is simply a matter of government policy and enthusiasm. When we hear that new solar installation produces 80 percent of a town's power requirements, a default shortcut is to think we can get rid of 80 percent of the fossil fuel consumption. Solar proponents do nothing to discourage that line of thinking, but it does not work that way at all.

THE WHOLE WORLD WANTS TO CONSUME LIKE WE DO TOO

It would be bad enough if all we had to worry about were our energy systems and how to switch them from a fossil fuel base to a renewable energy one. To do that alone is like getting every person in China to hold still for a minute to pose for a satellite photo.

But it actually gets much worse, because the big bad systems we think we can change away from in some reasonable timeframe are growing, even in parts of the world where they don't yet exist. This is why China continues to build coal fired power plants, and many regions of the world are investing in energy infrastructure as well. Even Saudi Arabia is investing billions in new petrochemical plants and refineries. Even many green developments are only switching one type of fossil fuels for another; China is building huge natural gas transportation systems to replace coal usage where possible. That is certainly a good idea, and it will help with greenhouse gas emissions, but it's yet another fossil fuel system being constructed.

Making matters worse is the global standard of living. It is, to be extremely polite, uneven. The life quality of the top decile of countries has about as much in common with the bottom decile as does a Rolls Royce with a rickshaw.

It would be hard enough to halt greenhouse gas (GHG) emissions and rising CO_2 even if we froze everyone's living standards where they are; systems cannot be adapted in any short time frame. That would be a monumental challenge. When we factor in the growth trajectories of less developed countries, however, the challenge is overwhelming. In chapter 5, we looked at how many vehicles would be added in the world if even a fraction attained western living standards, and that's just vehicles. What about everything else?

If we were all Buddhist monks, for whom being a materialist is having your own bowl, we would perhaps be able to rest more easily about our environmental footprint (of course, if we were all Buddhist monks, we would rest more easily by definition). But we're not. By the time we learn to speak, we've outgrown more than a few bowls already, plus a sizeable amount of other consumer goods as well. From there, impossible as it sounds, our consumptive habits actually grow.

It is hard to wrap our heads around the quantity of stuff that we consume. We can single out a simple item and try to picture that, and it will almost fry our mind. Consider shirts (or blouses, or whatever they are called on your gender matrix). We all hit a point where we need to clean out our closet and decide to purge a bunch, in a pile on the floor usually. Now, imagine what that pile looks like for a town. Or a city. Or a country. At some point, it loses meaning, because we can't picture that, and that is just what we throw out routinely.

Luckily, for the fossil fuels discussion, we don't have to try to make our brains do something it clearly doesn't want to. We can just jump ahead in the discussion to a place that's relevant and useful.

Consider how difficult it would be to live without all the consumer items we use every day. Without any of them. To take the extreme case, that's what it would mean if fossil fuel production was halted rapidly. This simply needs to be the starting point of where the discussion begins on the best way to get off fossil fuels. We have to consider what our life looks like without them, because we can't paint an understandable picture of how much we use of fossil fuels. We can't see that for ourselves, never mind the world.

At the same time, we need to wed this vision with what the opponents of fossil fuels are trying to do to save the planet. If we start with the premise that getting off fossil fuels means disrupting petroleum supply, which is the current theme of the environmental movement's program, we can see that there will be no progress on any front, other than headline-grabbing confrontation. The environment won't be saved either. Consumption is the problem, and the key to the solution.

But just as consumption is the true root of the problem, we should also expect the fossil fuel industry to be pulling in the same direction. It is obvious that increasing fossil fuel consumption is good for the companies and jurisdictions that provide them, but that doesn't mean that the industry can't be improved. It's one thing to say we can't live without fossil fuels at present; it's another to say we can't do a better job of providing and stewarding them.

17.
The petroleum industry can do better

WANTED: CULTURAL REVOLUTION

Deep in the woods of what is known as the Alberta Deep Basin, a geographical region with an enormous variety of hydrocarbon-bearing zones far below the surface, an oil patch worker grabs his trusty shovel and heads out to a mound of dirt behind the natural gas processing plant where he works. We'll call him Wes, because that is his name, and today he is not on the hunt for hydrocarbons. Wes and his shovel are going gardening. For hardened and calloused operations personnel, hobbies tend towards hunting, fishing and ATV'ing; gardening is like wearing work gloves with a floral motif. The laughter of his coworkers rolls off his shoulders. Perhaps if the crop pans out, a potato cannon will join the shovel in the tool shed. Then we'll see who's laughing.

Wes is testing an idea. When the gas processing plant where he works was constructed, rich, beautiful soil was scraped off, piled up, and forgotten. But Wes didn't forget it, and he decided to see if he could grow vegetables on that site. The first year was a test, and it was a success. The potatoes grew. Wes said, "No bears dug it up, and most importantly no cats sh_t in it." He is a practical man. Not only that, he sees a bigger and better picture as well: his crop is headed for the food bank.

Not only is Wes practical, but he is a doer. Some people would look at that pile of dirt year after year and think, "Wow, that is one big pile of dirt." Others see opportunity. Wes is one of a stratum of people that are responsible for much of the progress we take for granted daily, all based on an attitude of, "What the hell, let's see if it works."

THE END OF FOSSIL FUEL INSANITY

Wes's vision extends beyond potatoes, and even past vegetables altogether. He also had another brilliant idea, which would have been a lot more significant than a bucket of spuds. He previously worked at a giant natural gas plant that processed tens of thousands of barrels of water per day, which was produced with the natural gas and was being reinjected into the earth. The water came out of the ground at a toasty warm temperature (60 degrees Celsius). In many facets of industry, free heat can be a good thing, and Wes had an idea to use the waste heat to generate power. By his calculations, with some plumbing and hardware, the system could have generated about half a megawatt (enough for 375 homes[1]), free of charge. He brought the idea forward and was promptly shot down by some hyper-rational engineer on a ruthless cost-benefit analysis.

That last point is not a cheap shot at the engineer; he most likely was simply doing his job of spending capital wisely and only on projects that met the acceptable corporate rate of return. That has been life in the oil patch for as long as anyone can remember.

We can see of course that this rigid attitude has a downside. People like Wes learn from these experiences that they shouldn't bother. Eventually, all that's left is a person, a plan, and a shovel. No one spends, no one gets hurt. Capital discipline is good, of course, but it often leaves no room for anything that hints at progressive projects. All we're left with is the near-free initiatives, because Wes doesn't need anyone's permission to plant a potato. His other good ideas die on the vine.

One fantastic thing about the energy business is that people like Wes are common. It is an entrepreneurial industry filled with men and women of initiative. They are concerned not just for their pay, but for the betterment of the community, for the proper usage of resources, and because they're natural builders and doers.

This is the same for most industries, of course; these types of people are everywhere. Their ideas are also shot down everywhere. But we are here to talk about the fossil fuel industry, which is perplexingly and simultaneously both open to cutting edge technology and scornful of new ideas like planting potatoes in unlikely places, or using waste heat to power something out of the ordinary.

THE PETROLEUM INDUSTRY CAN DO BETTER

One could argue that in a cyclical industry like fossil fuels, there is no room for such experimentation, or for investments that don't match the returns of the best plays being developed. In the absence of any societal concerns, that might once have been the case, but these days, stakeholders would like to see a little more.

We can see that progress in these areas is possible though, through an example like Wes's that came into reality. Oddly enough, it involves the vegetable world as well, and even odder still, it occurred in France. It was a European oil/tomato* venture. I'm not talking olive oil, but about real black-gold hydrocarbons.

A Canadian oil company, Vermilion Energy, has production operations in France. That's right, France produces oil. The similarities with Texas never seem to end. Vermilion has a battery site located near Paris, and in conjunction with local agricultural engineers, a system was devised that uses the heat from water produced along with the oil (naturally heated to 60 degrees Celsius) to heat greenhouses. The water that comes out of the ground commingles with the oil flows into a heat exchanger, where a completely separate (and well-sealed, presumably) water system is heated, which then carries heat to the greenhouse. The separated oily water that comes up out of the well is combined with ultra-cheap wine and sold to North Americans under the banner of an exciting new variety of grape (okay, I made that last part up; it is reinjected into the ground).

The result of this program is 6 thousand tonnes of tomatoes per year that can be shoveled directly into Parisian gullets, 150 jobs created, tax revenue for local governments, and a staggering amount of good will and good press.

Ultimately, this is a good story for Vermilion, its shareholders, the greenhouse employees, the environment, and Paris itself. It is not a good story for the rest of the petroleum industry, for the same reason it's not a good story when a little old lady thwarts a mugger by thrashing him with her purse as a group of able bodied people stand by and watch. In isolation, we would look at the lady and think that was a pretty cool story, but when

---

* To those pedantic nitpickers who will leap from their seats in outrage that the tomato is incorrectly called a vegetable and not a fruit, please note that the US Supreme Court in the case of Nix v. Hedden of 1893 deemed that, under US customs regulations, a tomato should be classified as a vegetable. Which reference is, now that I think about it, pretty pedantic too. So never mind, call them whatever you want.

we know there were a dozen people watching and doing nothing, it's less exciting. This is a roundabout way of saying that France has about two oil fields and North America has a billion—why did it happen there and not here? Why didn't it happen here a thousand times over? Why does Canada import one vegetable from the United States when we have enough waste heat to make enough ratatouille to feed China?

We can't say we don't think of such things, because people like Wes certainly did. It is a bit ironic and should be thought-provoking that, in the heart of the oil patch, the only food chain that can be derived from waste resources is a solitary figure digging up potatoes, while in a country with no preconceived notions of what is possible, a greenhouse can be built.

STAKEHOLDER RELATIONS NEED TO BE MORE THAN A PLATITUDE

Some aspects of the petroleum business are definitely changing. The supermajors are changing because the world is forcing them to. Governments no longer welcome them with open arms wherever they go. In fact, many governments are going the other way, such as when certain jurisdictions sue big oil for their complicity in climate change.

The rest of the energy business is changing as well, but there is still much work to be done for the business to approach standards that are anywhere near cutting edge. The old attitude of big oil, that the path will be cleared for them to produce the resources the world needs, still exists down the food chain to a certain extent. We unfortunately see instances where not just the industry but regulators have not kept up with the spirit of the phrase "stakeholder relations" that is so easily tossed around.

As an example, an Alberta woman who we will refer to as Diana, because that is her name also, illustrates the issue very well. I first came across Diana when inadvertently reading a comment section annotating a news article. Comment sections tend to be a revolting world, populated by unfiltered and short-tempered crackpots who operate under cloak of anonymity and react reflexively and viciously to any notion that is not in absolute laminar flow with their hair-triggered frantic brains. Some seem to be working for certain organizations or, if not, are simply unhinged enough to spend their days (and apparently much of their nights) hunting down anyone that doesn't share their views and giving them a simulated and enthusiastic beating.

At any rate, once in this netherworld, I noticed a person who was quite active and not in a particularly pacifist mood. She was jousting with combatants, as often happens, but her stance was peculiar. Incredibly well researched and nimble of debate, she carried the frustration of the best of the combatants with two feet firmly on the ground, like a warrior calmly batting away waves of crazed attackers with her sword. In the instance I read, she eviscerated a rude self-described CEO, one with far more inside knowledge of the oil business than she had, who kept getting angrier and angrier at her comments. Diana never lost her cool or swayed from her argument. The enraged anonymous CEO lashed out, whereupon Diana simply read back quotes directly from the guy's professional code of ethics, and he disappeared.

I also noticed that Diana had asked me a pointed question with respect to one of the posts I had written. It was not nasty or provocative, but simply direct and forthright. Her comment directed to me was that yes, we get it, why do you keep belaboring the same point (I had had several consecutive posts dealing with the same topic) when there are constructive things to write about? What intrigued me was that I had been thinking the same thing; the topic of the season had been a pipeline controversy that I'd been yammering away about for several months to the point where I was even annoying myself.

I took the liberty of contacting Diana, not sure what to expect, because her online presence is formidable—assertive, unafraid (especially so considering she posts under her own name), and well researched. I certainly didn't agree with everything she posted, but her points were at least intelligent enough that I could weigh them against my own, as opposed to the vitriolic nonsense that simply makes me want to crawl under the covers and disavow the human race. I also did a bit of research and found out what had turned Diana into the fighting machine that she was.

As it turns out, she is a mother and vegetable raiser and not at all a wingnut. What happened to her was a grueling multi-year battle with a sequence of energy companies that drilled wells near her family's farm, and the company breached several rules with respect to operations. Had it all been addressed by the company properly and up front, no one may have ever heard of Diana again (well, who knows, maybe she is destined to develop world class piano skills or something, how would I know). Instead, she was faced with a company who viewed her as an irritant, and subsequently a regulator who found the same.

She would not have been an irritant if the standards in place had been adhered to, and if someone would have made a sincere effort to deal with the problem. Such was not the case, and as she faced obstruction after obstruction, instead of folding, she grew more resolute. She also demanded answers, and instead of receiving them, the counterparties got tired and tried to sweep her under the rug.

What followed for her makes gruesome reading. An annoyed company and a regulator who simply wanted everything to go away both became ever more aggressive towards Diana, who was simply holding out for acceptable explanations or responsibility for things that were clearly happening to her, her family, and her farm. The company that drilled the wells near her farm dubbed her "irrational" in an email to a PR consultant they hired. As you can imagine, the strategy that this media relations expert and the evasive company cooked up was nothing short of flamboyantly ill-conceived. The provincial regulator put out a report documenting every one of Diana's emails and when they arrived, the whole of which appears to be a defensive "we did our job" kind of document that is a bit embarrassing to read. The report points out things like that there may have been venting of sour gas that may have been offensive, but it has stopped now so there's nothing we can do and how can that be a problem?

This incident brings to mind several other similar situations that have happened in the oil patch over the past few years, but ones that played out on a completely different trajectory. There were several instances where petroleum companies attempted to drill wells in more affluent parts of the province, for example just south of Calgary, in the beautiful foothills that attract wealth. Here, when planning to drill wells, oil companies were met not with farmers with questions, but by walls of lawyers, courtesy of the wealthy locals. Having made their fortunes in the oil business themselves, these groups knew how things really work, and launched their protests before the wells were drilled. They left no chance for future debates with regulators about whether any venting was going on, and nominated attractive people to speak to the media about their concerns. The contrast between the two situations is not a pretty one.

Diana's situation need not have ever gotten to that point, but the fact that it did illustrates another example of a problem that the petroleum industry will face: that the world is indeed changing, and that more is expected than in decades gone by. As with pipeline companies and big oil, the whole business needs to understand the fight it is in. There is a difference between meeting acceptable regulations and doing the right thing. The latter is the way the world is moving.

These episodes might seem minor in the big scheme of things, and perhaps they were once upon a time before the internet and all it has brought. But for an industry on its heels and playing defense, such episodes illustrate the explosive nature of handling these things poorly or in a dismissive manner that worked in the past but certainly does not now.

A single incident like Diana's, provoked by a recalcitrant company and a regulator unprepared to deal with the new world, leads to a story that spread far and wide. It blackened many eyes and forced a vegetable-growing mother to become an amateur lawyer, historian, and regulation-junkie combatant. No one wins. Instances like these were once an irritant that no one ever heard of beyond the farmer's immediate circle and perhaps the local newspaper, but the internet has smashed that cozy world forever. It may seem unjust to some, but a single publicized incident like this is now part of the worldwide reference library on bad corporate behavior. For better or worse, there is a lesson here that absolutely must be learned and acted upon if the petroleum industry wants to earn global respect, and this wasn't the way to do it.

It is a lesson in maturity and being a stand-up citizen. It's not okay to act appropriately most of the time, particularly when, as the oil patch is in Canada anyway, developing the resources that belong to the government/ province. It's not okay to go on the offensive to discredit a stakeholder with valid questions. There are no rewards in life for not doing something, no fanfare for not punching someone in the head when they are being a total jerk. For an energy company, there are not always rewards for doing the right thing, for erring on the side of those whom you work amongst. But there are colossal repercussions for punching someone in the head, even just once. There are repercussions you can't even dream about, because someone in Poland or Madagascar or Togo can read about this woman's experience and propagate the word, and before you know it, Canada has a reputation for something. The point here isn't that companies should do the right thing for publicity reasons, but that the cost of doing the wrong thing is a lot more substantial than it used to be.

Many industries have cottoned on to this, as can be witnessed in any establishment that has been dipped in a vat of modern customer service theory. This plasticized attention and devotion to your satisfaction is almost embarrassing at times, like the forced fake smiles of people making minimum wage, grinning like they are ecstatic to serve you, and the frenzy to fix something if it's done incorrectly. There is a palpable fear in the service industry to never set a foot wrong, to never get bad reviews on a review site. You get the feeling that their life almost depends on your satisfaction, and it might just well be the case if jobs are scarce and the qualifications are modest.

The whole sad customer-focus scene sometimes seems stupid and over the top, because we can often see through it. But there is a hard, sharp, real reason that the world has gone this way. We can mock it and live by our own code of yesterday, or be run over. The petroleum industry is correct to believe that the world can't live without it at present, but it also needs to remember that these issues fester and grow out there in the world, and that there are many people just itching for a chance to help the fossil fuel business hasten its own demise.

TIMES ARE CHANGING, EVEN IN THE OIL PATCH

It is hard for people who've never worked in the oil patch to see how far it has come in short order. It has a long way to go, but to think it hasn't changed is profoundly untrue.

For one thing, it is now tragically boring. When I started in the business in the 1990s, my first office was on the ninth floor of an office tower in downtown Calgary. I was lucky enough to get an office with a window, and frankly I felt like I'd won the lottery because my office faced the once-legendary 3rd Avenue strip, and from 11 am on, I had the privilege of watching the hookers work that little stretch of pavement. As a boy off the farm, it was quite beyond anything I'd seen. It was a good thing I worked for a major oil company, because on nice days I'm afraid my productivity was probably nothing to write home about. The ladies were surprisingly busy from lunchtime on.

The subsequent history of that strip is reflective of how the business has changed too. The prostitutes liked that stretch of road because it was right outside the French Maid, a "gentleman's club," the description of which always made people double over with laughter. Today, a mighty glass tower occupies that space, and the modern, open, airy, granite lobby does not contain a single trace of where the stage once stood. The new building's owners were apparently unable to incorporate into the ultra-minimalist décor even a commemorative strip pole.

Having read this far, perhaps you'd appreciate a better reference point than to read rambling about prostitutes. Here is a concrete example then, which will be dealt with more extensively in a subsequent chapter.

In 1989, the Exxon Valdez oil tanker crashed into a rock off the coast of Alaska. There were several reasons why it happened, many of them attributable to Exxon itself. As the matter worked its way through the courts, Exxon put on a textbook display of the ex-colonial attitude discussed in an earlier chapter, and fought tooth and nail against any settlements. They argued and appealed and delayed right to the US Supreme Court, and reached final settlements in 2008, some nineteen years after the incident.

In 2010, BP had an enormous petroleum release in the Gulf of Mexico. Recall that BP was one of the original Seven Sisters, so subject to the same world view as Exxon. But times had changed, and the world's tolerance for nineteen-year court battles over major pollution incidents had evaporated. BP's response had some of the hallmarks of the old-school mentality, with a CEO publicly lamenting how he wanted his life back because he was tired of dealing with the mess. However, despite some relatively minor old-school thinking, most of the matter was settled within five years of the spill, including massive payments into a fund to handle future claims. When one considers that the Exxon Valdez spill impacted raw wilderness in an area of the world with few people, and that the BP spill impacted prime tourist destinations and the pristine blue waters of the Gulf of Mexico, the speed with which BP settled things, at a cost many, many times what Exxon spent, is indicative of the realization among oil companies that the world has changed.

DESPERATELY NEEDED: A GROUND-UP GROWN-UP ATTEMPT AT OPENNESS, HONESTY, AND TRYING NOT TO SOUND FIFTY YEARS OUT OF DATE

It is important to face the reality that the fossil fuel industry is under attack. The attacks may be over the top, or outdated, or exaggerated, or sometimes outright false, but most of the world doesn't really care. No one ever has, or ever will, shed a tear for ExxonMobil, or for pretty much any energy company. Much of the world sees the attacks on the fossil fuel industry as some sort of cosmic comeuppance for decades of big oil dominance and for watching unhindered development lay waste to many natural landscapes. There were undoubtedly many eyebrow-raising corporate tactics and situations in decades gone by, and even if they no longer occur today (or not nearly as frequently anyway) the memories linger. Fairness has nothing to do with it; it may not be fair to expect small independent companies to bend over backwards to make stakeholders happy, but neither was it fair when some oil company came to a developing country and cut a deal with a dictator and evicted a tribe off its land.

What is relevant is that the fossil fuel industry has come a long way, but has a way to go yet. Corporate communications remain abysmal, arrogant, and half-hearted. If fossil fuel companies refuse to acknowledge the world's concern with global warming (note the wording there: if they refuse to acknowledge *the world's concern*), the least they should be doing is emphasizing environmental best practices, reducing corporate footprints, and in general acting in a manner that is in lockstep with the world's expectations in this day and age. They should be helping Wes plant potatoes, and maybe adding a few greenhouses alongside. They should adopt the mantra of many businesses that rely on good faith relationships with the public, and make certain that people like Diana are heard and treated fairly and worked with, not against, until acceptable solutions are met. They should recognize that what is acceptable to an engineer with the book of regulations as his guidepost may not always be enough.

THE PETROLEUM INDUSTRY CAN DO BETTER

The attitude of some of the old guard will be that things shouldn't have to be that way, that it's ridiculous to cater to the extremes or accept that they should step up to do something outside the norm in the name of a social license. Some will look at their sterling credit ratings and stable cash-flow-generating businesses, such as those in energy infrastructure, and decide that their communications programs are just fine the way they are.

All those people can stand their ground, and on principle can continue running their businesses exactly the way they want. As long as they do though, they need to be cognizant of the fact that they're no different than some cranky old Kansas farmer choosing to ride out a tornado warning by sitting in his living room watching TV because no one tells him what to do. Enjoy the ride.

18.
Oil spills and other nastiness

Globalization has brought many benefits: international trade, international travel, and on a more personal level, the knowledge of how finite our planet is. A hundred and fifty years ago, an adventure novel called *Around the World in Eighty Days* was published. The title refers not just to the adventure, but a bet as to whether or not the protagonists could actually achieve it. The journey is littered with trials and tribulations of the time, such as learning from a British newspaper that a new rail line had been constructed across India, and upon arriving, finding out that it was minimal and of little help in crossing the country.

Today, we can be out for a walk in the park and pull up on our phones a webcam of some particular train station in India and see what's happening at that very minute.

This isn't news, obviously, but the point here, and of the whole book, is to consider the systems that make it all possible, what is required to make all those systems work, and the size of the task in changing them.

All this magic of modern life involves a mind-numbing array of chemicals, industrial processes, and transportation networks. We see the magic, but we do not see what's required to make it happen. A magician "cuts" a helper in half with the aid of a trick box that is lugged around from show to show. An iPhone is a chemical stew needing 75 of the 118 elements of the periodic table. There are no illusions in that trick, but it is nearly magical that it actually happens.

Those cellphone components don't all come from anywhere handy like New Jersey. They don't come from anywhere you've ever been, unless you have an extremely eccentric bucket list. You will not find yttrium, terbium, europium, or gadolinium at the local mall or by panning in a stream (all these are used in iPhones), although don't be surprised if you find them in your pyro maniacal son's bedroom. Best not to ask. They don't all come from Shenzen either, or wherever in China cellphones come together. The ingredients come from every corner of the earth. For example, one of the biggest repositories for cobalt, a necessary ingredient in rechargeable batteries, is in the Democratic Republic of the Congo. From the mine in the DR Congo to China involves, as you can imagine, much transportation.

Cellphones are one example. Now add in everything else you use daily and consider all the ingredients required Now, think about that times 7 billion people. Now, think about how all those things got to where they need to be. A perfect system would have some sort of magical Star Trek-style transporter beam, but our damned engineers can't even make anything levitate. Thus, we rely overwhelmingly on fossil fuels to move everything.

And because it is an imperfect system, guess what?

SH_T DOES INDEED HAPPEN – INDUSTRIAL ACCIDENTS ARE AS HARD TO ERADICATE AS RATS

Have you ever worked with machinery, such as in a manufacturing plant or construction site? If you have, you know that moving parts mean imperfections. Hoses break, people crash into things, machines leak, valves fail. Weather plays a part, as does human fatigue and metal fatigue. The more controlled and stable the environment or process is, the safer it is in general, but many industrial activities are on the other end of the spectrum. In the oil patch, a missing-digit hand-wave has historically been a typical greeting of a rig worker, whereas an accountant losing a finger in the line of duty is a rare event indeed—that would be the mother of all paper cuts.

Much of the world's fossil fuel transportation system and infrastructure fall into the rough and tumble end of the spectrum, as opposed to, say, dentistry. A dentist operates in a perfect environment with sterilized instruments and exactly the right lighting and mechanical conditions. The tools are of extreme precision, the power supply is never interrupted, and he/she never has to change oil in the machine or check the coolant level. Even with the perfection of the environment, the dentist wisely avoids the risk of operating in a fatigued state by shutting down five hours early, when the coffers are full, because there are new titanium irons to break in down at the golf club, and new Lamborghinis don't just pick themselves out.

Out in the real world, it's too hot or too cold, the sun is in your eyes, you're tired because it's been a long day, some equipment needs inspecting, a pipe breaks because the ground shifted, or a truck breaks its suspension after hitting a badger hole. Or a million other things that can and do happen every day.

The worst incidents in the world have been a combination of a few relatively small bad things happening in one day, each multiplying the effects of the other. If there are a million unforeseen things that happen every day, there are a billion close calls. The sheer randomness of these industrial processes means that sometimes things line up, like selecting the right numbers for a lottery ticket but with a kick in the teeth for a prize.

This isn't an excuse for industrial accidents; they are simply a fact of life as we know it. Businesses and governments strive mightily to make work places safe, and employees do also, but perfection is unattainable because there are too many random events out there, and sometimes they just line up against you. You might have a thousand close calls in your life and none of the real deal, or you might have ten significant accidents and none of them your fault.

THAT DOESN'T MEAN ALL ACCIDENTS ARE THE SAME

At some point today, an overly-tired worker will drive a truck into something in a way that may or may not be really funny to see. The incident may or may not have serious consequences, and tomorrow, the odds of it happening again are extremely high, somewhere in the world. It is pointless to expect perfection from the millions of labourers out there who perform these tasks daily.

That's one example. Some have far bigger repercussions than a truck bumping into something. Consider the malfunction of a simple O-ring seal on the Space Shuttle Challenger in 1986, the one that blew up shortly after departure. The seals were made of a type of rubber that became brittle at the unusually cold temperatures on the morning of the launch. The results were, to put it mildly, not funny at all. The company that created the seal was not a roving band of idiots, and the failure was found to be the result of a number of individually minor shortcomings, such as a weird organizational culture at NASA. Many organizational cultures are weird but don't end in world-famous explosions. It was caused by the culture, plus substandard decision-making processes, plus a lack of communication, plus an unusually cold morning.

Yet another example, closer to the world of energy, is BP's massive oil spill in the Gulf of Mexico in 2010. Eleven workers were killed and four million barrels of oil spilled into the blue waters of the Gulf.[1] This incident will be discussed later in the chapter, but it is worth noting at this point that eight different circumstances or shortcomings existed simultaneously that in combination led to the underwater blowout. None were huge problems independently, and neither were they necessarily fatal. It was simply a matter of probability that enough of these would convene at the same time and place to cause a disaster.

BP and NASA are not bumbling fools with third-rate equipment that is scarred by countless run-ins with immovable objects by careless operators. BP has been a pioneer in oil and gas exploration for decades, including offshore drilling, and NASA is, well, NASA. They put a man on the moon.

WAIT A MINUTE, IS THIS SOME SORT OF EXCULPATION OF PIPELINES AND ALL THEIR LEAKS?

It could be construed that way, depending on how you look at it. There are about three ways, it seems.

A pipeline opponent would say yes, it is, that because accidents happen all the time is no excuse to put clean waters or the environment at risk. One might say water is kind of important, if one is prone to understatement. This group would put forth an argument something like Nassim Taleb's Black Swan analogy, whereby it's not the frequency of something that is important, but the severity and the unexpectedness. A hundred small spills at pipeline pumping stations are not the same thing at all as the same volume dumped at once into a seaside community.

A pipeline proponent would say incidents are unacceptable and no, that's not part of the pipeline story at all. They would maintain, per key message guidelines, that they have safe operations, and generally don't have much more to say on the topic. That is not an insensible position. It is, I suppose, a poor business practice to shrug when talking about one's own pipeline incidents and say, "Well, sh_t happens." Regardless, the position of the pipeline company would be that their operations are safe and that they take the matter of safety seriously. And from what I've seen, they really do.

What then would be the third way of looking at this? It would be from the systems perspective. It would be to set aside whether pipelines or evil, or whether they are 100 percent safe, or whether they promote global warming, or whether they are causing an epidemic of burrowing gopher concussions. What is of significance is that 7 billion people are currently consuming 100 million barrels of oil per day. We have a choice: we can stop all those people from consuming all that oil, somehow, and face the repercussions of that, or we can do the job to the best of our human ability until some distant day when oil isn't as dominant. The impact of the former can only be speculation; if we're going to slash oil consumption immediately to the point that pipelines are only marginally required, who will be first to go without? Any takers? I didn't think so, because that door is always wide open, and no one chooses to go through it. No one. The latter option, of doing the job to the best of our abilities, is what we need to focus on.

Having said that, we currently live in a world, at least in North America, where pipelines and oil transportation systems are vilified. They are ubiquitous, silent, and effective, but vilified. We will look at distribution systems in the next chapter, but for now let's focus on how we got to the present state of animosity.

THREE EVENTS REDEFINED OIL TRANSPORTATION

The world of petroleum transportation, with hundreds of thousands of miles of pipelines and countless connections and valves and trucks and tankers, is now encapsulated by three names: Exxon Valdez, BP Gulf of Mexico, and Enbridge Michigan.

It may not seem fair that such broad facets of industry can be doomed and defined by singular events, but it happens. It has happened in other industries. In 1926, a couple of scientists developed a new form of rubber, and built a business over the next sixty years around it. Few people had ever heard of the new rubber, but it was useful industrially, including being used in seals for rockets. Things were just fine until 1986 when the aforementioned space shuttle disaster thrust the name Morton Thiokol into worldwide infamy. The company name became synonymous with disaster, which is a situation infinitely different from the one that existed thirty seconds before the space shuttle explosion. The world changed forever, at least as far as Morton Thiokol was concerned.

The same phenomenon has befallen petroleum transportation. The entire business of moving petroleum around has been redefined by these three incidents. Opposition to pipeline projects or other energy infrastructure usually includes at least one reference to spills, and when they happened, the incidents dominated media headlines for months. Sometimes the actions of the perpetrators ensured the incidents stayed in the news for years, as with Exxon's Valdez spill. That event is, thirty years later, still a common reference point for those who want to be rid of fossil fuels.

Despite the fact that these three industry-defining incidents are all "oil spills," it is worth delving into the circumstances of each a bit more. The incidents have almost nothing in common except that oil went into water, and that multiple coinciding factors led to the disasters. It's worth taking a few minutes to look at all three to see if the circumstances that led to each has any bearing on what we should be watching out for next.

EXXON VALDEZ TANKER SPILL

No, he wasn't drunk.

That was the myth that surrounded the Exxon Valdez tanker incident of 1989. Rumour had it that the captain was under the influence as he piloted the oil tanker at night. It makes a really good story, like the craziest drunk-driving story you'll ever read. But it's not true.

Everyone, especially Exxon, probably wished it were something as simple as that, and in fact, according to Wikipedia,[2] the company did sort of throw him under the bus by blaming him. As with many of these mega problems, the truth seems far more mundane. That's not to say the outcome was mundane; that's not the case at all, but each component of the disaster, in isolation, seems not particularly relevant.

Again, according to Wikipedia, multiple factors contributed to the incident, a familiar theme for these rare but cataclysmic incidents. Factors involved here included an overworked/under-slept crew, a radar system that was not functioning properly, and that the ship was travelling outside normal sea lanes to avoid small icebergs.

The location itself contributed to the disaster. Alaska is not exactly New York City, and remote areas make it exceedingly difficult to get emergency response equipment to the site. Think about hundreds of miles of almost uninhabited, extremely rugged coastline that is difficult to access—roads aren't everywhere. Think about the difficulty in reaching the shore by water, in rough seas and in areas that are dangerous to approach. Those conditions, not sandy beaches with gentle waves, are what dominate much of the Alaskan coastline.

There is no need to go into a full analysis of the incident; I'm far too lazy and you can read better accounts of the entire fiasco elsewhere. What is relevant in the context of this discussion is the extent to which this spill helped define the world's attitude towards maritime oil transportation.

The fact that the Valdez spill is a defining event of water-borne oil transportation is unfortunate. Over 50 million barrels of oil per day move on waterways, which is a lot of ships and a lot of loading/unloading terminals. If you think about it, the entire Middle East, the North Sea, any offshore drilling, and any production from any small oil-exporting waterfront nation makes its way to a point of end use via water. Every day, thousands of these vessels, some not so big and some enormous, move about, loading and unloading, in stormy seas and calm ones, almost all without incident. There is no choice; it has to be a relatively safe method of transport because it is so critical to the system of energy movements that keeps us all alive every day.

BP GULF OF MEXICO

Moving on to the next of our filthy troika, the BP GoM incident finds us far away from the gloomy, desolately haunting beauty of the rugged Alaskan coast into the beautiful Gulf of Mexico. Blue skies, even bluer waters, and endless black blobs of foul smelling goo popping up on the beach. Ah, the summer of 2010.

From an environmental incident perspective, there's no difference between this and the Exxon Valdez other than magnitude. A petroleum process of one sort or another that went hopelessly wrong, with a resulting mega-mess in the water. That perspective is understandable but only on the assumption that one neither knows nor cares what our standard of living requires, and what sort of activity needs to happen every single day for us to live as we do. We need to therefore examine this incident in light of its place in the energy system, and see what we can learn from that.

The Exxon Valdez spill was the result of moving hydrocarbons from a production site to a refinery. The BP GoM incident was the result of trying to get the stuff out of the ground in the first place. That may seem like hair-splitting, but it's not, and it is important to understand just how different it is. Viewing the whole ordeal of finding, producing, processing, and transporting petroleum as a single process dangerously underestimates the complexity. As before and forever more, complexity and ingrained systems are susceptible to incident, and fiendishly hard to change.

With respect to finding and drilling for oil, a familiar refrain accompanies this symphony (or cacophony) of processes and moving parts: that the complexity is staggeringly huge. First off, drilling offshore is now commonplace but is an extremely challenging undertaking. Consider looking for petroleum on shore: you find a prospective underground location, find a suitable site to drill, prepare the site, and haul everything you need over there and get going. Drilling engineers would beat me over the head for that overly simplistic description, but based on current observations I can outrun most of them, so we'll work with that. Now imagine doing the same thing offshore, with wild seas and storms and no roads. There's no anything. Furthermore, with onshore drilling the rig is precisely located on a purpose-built pad; for offshore drilling the target is anywhere from tens of meters to over 3 kilometers (2 miles) below the surface. It can be like assembling a Swiss watch at the bottom of a tank that holds three feet of mud, while wearing gloves. If you are thinking how dumb it is to be searching for oil in such inhospitable environments, remember that you are the cause of that just like everyone else, every time you consume something or get on an airplane.

That deep-water drilling happens at all is a miracle, with all the things that can go wrong. There is probably a useful statistical analysis somewhere that would indicate that almost every job has a few unplanned occurrences, or policies that were not followed precisely, or some sort of equipment failure. Over a large enough sample size, some jobs will have none of these instances, some will have an average number, and every now and then one will have far more than average. That's how averages work. It will also happen, even more rarely but statistically it will most likely happen, that the jobs that have an aberrant number of malfunctions will have an aberrant number of malfunctions that exacerbate others. For example, one worker cutting his hand is an aberration, but what if two workers cut their hands? If they are completely independent incidents, then it doesn't matter that much, but suppose that the two people that cut their hands are the only ones whose job it is to man a safety switch or override valve or something. If they both go down, then the operation is exposed to a catastrophic incident.

The more job functions one includes in the sample size, the more likely one is to encounter such a situation. With the BP GoM incident, eight separate safety failures happened to occur simultaneously. The issues included two separate valve failures, which are rare in themselves, a poor cementing job, test data of this cement job that was misinterpreted, and finally a faulty blowout preventer. This latter piece, a vital safety mechanism, is designed specifically to prevent these incidents. It comes with its own safeguards including two backup systems, but it had a flat battery in one system and a defective switch in the other.

If any one of these failures had not occurred, the other seven may not have mattered. But that's how the world works sometimes. It happens to new moms trying to look after a baby and make dinner and do laundry and plan the next day's activities. Some days it goes perfectly, some days the dog throws up on the carpet, the dinner gets burnt and the baby gets a bowl of sour cream to eat. It happens everywhere. Sometimes a small malfunction happens in isolation, and sometimes six happen together and a major incident happens. That's not petroleum, that's life.

ENBRIDGE MICHIGAN

I can't begin to imagine what it was like to be the pilot of the Exxon Valdez, in part because I despise open water and refuse to get on anything larger than a kayak except a cruise ship, in which case I break both rules but stay drunk for eight days straight. I also can't imagine what it was like to be overseeing the drilling of a deep water well, because of the water thing again, and because I have never been remotely close to working with that sort of technology in a complicated and remote environment.

Concerning our third anti-star however, the Enbridge Michigan spill, I do have a tiny personal glimpse into how that sh_t-show unfolded. I've never worked in a control room (I've been in one and seen the fleet of monitors that resembles the New York skyline at night), I've never worked a shift on a pipeline that had 24/7 operations, and I've certainly never tried to restart a pipeline after several warnings that it was not a good idea. I therefore won't comment on the workplace habits/circumstances that led to what was described as a "complete breakdown of company safety measures" by employees who performed like "Keystone Kops" in trying to control it, according to the US National Transportation Safety Board in a July 9, 2012, report.[3]

I have, however, been witness to the sort of agonizingly slow underwater ballet that a response to a pipeline spill can be. I was once an observer to an oil spill clean-up simulation on a river. The simulation was planned weeks in advance. The crews showed up at the site, an hour outside a major city, at 7 am with the full armada already there and ready to go—response vehicles, support trucks, boats, floating booms, and all required equipment. The simulation was held at a boat launch, the only boat access point for miles (which is quite typical for larger rivers—boat launches are surprisingly far apart).

The successful test was a shocking eye opener. It took four full hours before the booms were effectively out on the water, the whole process occurring right before our eyes and within a stone's throw of the boat launch. Had the "spill" occurred a few miles from that comfortable site, which would in the real world have been by far the most likely circumstance, the time would have been even greater. In the event of an actual spill into the river, it would have taken some time to notice and locate, if the discovery was during the day. If the leak had occurred at night, and if it was small enough that the pressure controls didn't catch it, then it would have been until at least morning before it was sighted, but probably a lot longer. The odds are overwhelming that from the time of discovery to the time booms were in place would, in a real-world situation, been more than eight hours. If a river has a current speed of even three miles per hour, at least twenty-four

OIL SPILLS AND OTHER NASTINESS

miles of the waterway would have been contaminated. If the spill was in the winter? The room went silent when that topic came up in the debrief. I believe cyanide pills may have been handed out in a bit of forward planning, but I could be mistaken.

Does such a disturbing example mean we should stop using pipelines? No, because we can't. Not even close. There is no possible way to transport large quantities of oil and natural gas other than by pipeline. What this example, and the Enbridge Michigan real life one, tells us is that we need to be supremely careful about the maintenance and operation of pipelines. We do that for several reasons: first, because it is the right thing to do, and second, there is a not-small matter of potential liability cost.

THE COSTS THEY ARE A CHANGIN'

Here is one last tidbit on the topic of these three ugly siblings. Below are a few numbers that will give you some indication that "sh_t happens" is not on the list of acceptable answers for industry anymore. There are many reasons to work very hard to prevent incidents; here is one of the biggest.

These are the most recent cost tallies for the big three:

Exxon Valdez spill, which happened in 1989 - $4.3 billion[4]

Enbridge Michigan, 2010 - $1.2 billion[5]

BP GoM, 2010 – $60 billion[6]

You will notice a disparity there, particularly between the Exxon Valdez and the BP GoM. Both were massive spills, and the GoM incident was in a more populous region, but the magnitude in cost difference is staggering. The BP tally includes a $20 billion settlement with the US Justice Department. These are the modern cost exposures of a significant oil spill.

The Enbridge Michigan spill, whose costs are not remotely in the same ballpark as the others, is in some ways the most astonishing. The Exxon Valdez spill released an estimated 260 thousand barrels directly into the ocean, and the BP GoM release was a staggering 3 million barrels, also into the ocean, which helps explain the costs. The Michigan spill, however, was into a creek, and the release was less than 30 thousand barrels. The entire creek was accessible by equipment, was not remote, and no storms or high seas or even tiny waves were involved. Yet the costs still spiraled into a ten-figure sum.

BUT DO NOT THINK FOR A SECOND THAT ANYONE IS COMPLACENT ABOUT OIL SPILLS

The above situations, and the discussion, might sound like a resigned and lazy conclusion has been reached herein that there's not much that can be done; sh_t happens. Au contraire.

Sh_t does indeed happen, but that doesn't mean we accept it. For anyone who thinks the whole subject of spills is being taken cavalierly in the halls of big or little oil, think again. No one cares more about spills than oil drillers, pipeliners, or transporters.

In fact, the insurance industry is getting cold feet about insuring against possible spills, with fewer and fewer companies willing to offer blanket coverage. You can be sure that that development is causing a few sleepless nights around the world.

With that sobering summary, it's time to move on to analyze options with respect to moving energy. Remember, it has to be moved, it will always have to be moved, as long as 7 billion people need it, and as long as they refuse to all live right at the energy source. This anti-nomadic truculence is incredibly expensive, and it doesn't matter whether the energy source is fossil fuels, hydroelectric dams, or solar panels. Energy has to move.

19.
Transporting energy – options and learnings

Growing up on a farm in a rural area, one of life's great nuisances was dusty unpaved roads. Gravel roads dry out quickly after rain, and the rocks and gravel get pulverized by traffic to the finest dust imaginable that rises up behind a moving vehicle in plumes that can be seen for a mile. It was miserable driving on these roads, because the dust infiltrated everything, and meeting another car meant being suspended in (and inhaling) those particulates for what seemed like an eternity. It was miserable living close to one of those roads also, because every passing vehicle created similar dense, choking clouds of dust, which wafted across the yard.

A friend's uncle worked for the provincial power company, the monopoly supplier. He lived on a farm and was able to procure a seemingly endless supply of PCBs, or polychlorinated biphenyls, from work. This chemical is now unused and considered extremely toxic and damaging to the environment, and any spills are treated as major environmental problems. But in those days, PCBs were used in power transformers as an insulator, and when disposed of typically went to landfills. My friend's uncle showed typical rural ingenuity by intercepting the dump-bound loads, bringing the stuff home, and spreading it on the road in front of his yard to win the war against dust. We called him "lucky".

That demented bit of history isn't presented to promote the idea; it is an indicator of how far we have come with protecting the environment from contamination. Furthermore, with globalization, we are now able to use our findings about the dangers of these chemicals in the environment and pass those learnings around the world in hopes that developing countries don't let their farmers get too carried away either.

We know we can't treat the environment poorly anymore. We're more aware of how chemicals can affect ecosystems, and we have developed proper disposal schemes for many industrial products. We know how to make these chains safer through better management.

On the other hand, we need to be able to segregate events that are now unacceptable, such as dumping hugely toxic waste on roads, from events that happen in transportation chains that we cannot live without.

There is also another level of understanding needed. If a plane crashes for some unknown reason, no one, *no one*, proposes banning air travel, never mind protesting it. There is therefore a continuum of acceptability, of sorts. Somewhere between dumping toxic waste on roads and aeronautical disasters lie petroleum transportation incidents.

We think we *know* the following about energy transportation: oil tankers are dangerous, as proved by the Exxon Valdez; oil pipelines are dangerous, as proved by Enbridge Michigan; and oil drilling is dangerous, as proved by the BP GoM incident. Stepping further afield, we *know* nuclear power is dangerous because of, chronologically, Three Mile Island, Chernobyl, and Fukushima.

These things we *know* are not always helpful. We may know something is potentially dangerous, but we conflate potentially dangerous with unacceptably risky, and wind up in a real predicament, because we don't put in the effort to examine as carefully the dangers of the next best alternative.

We look at dangers only when we can comprehend the potential outcome, which usually means either we've seen it on TV, or when someone has gone to great lengths to spell it out for us.

Analyzing these dangers in isolation does us no good whatsoever, because they create a disconnection between what appears to be too dangerous to have around (but may not be), and the building blocks that are essential for our way life. We can't separate the two. If we want the lifestyle, we have to power it, and today—and for next decade, and the one after that—this means moving fossil fuels. A further face-paling scenario to contemplate is that the rest of the world wants the west's standard of living too, and sitting on the couch in our underwear eating home-delivered pizza and drinking

Coke is not the best pulpit from which to lecture the rest of the world that they need to make do with an ultra-spartan lifestyle.

Let's look at various transportation methods through a different lens. We don't need to pretend the big three oil/water disasters didn't happen, but we need to put aside the default reflexes that we've learned to associate with them.

TANKERS

There are many unsung heroes around. Some provide interesting stories that can make you feel glad to be alive or feel a bit of an inner glow, like when you read about a really good teacher that leaves behind an army of students saying how much better their lives are, or a business that donates end of day leftovers to a food bank. You can find all sorts of those stories all over the web and I encourage you to do so because, well, why the hell not?

In that spirit, what do you think about dump trucks? Or delivery trucks? I'm guessing you don't think about them at all, and if you do it's because they are in your way in traffic and are irritating, or you happen to catch a look at a driver who may or may not look like he just simultaneously rolled out of bed, prison, and a hangover at the same time, and you shudder involuntarily and speed away.

Imagine what your world would look like without those unaesthetic utilitarian boxes on wheels and their unsung drivers. Some haul garbage away, which definitely makes your day better. Other trucks distribute everything else we purchase. They bring fresh food, and they circulate all sorts of goods you would lose your mind without. Sometimes they are a nuisance in traffic or are driven poorly, and at those times, we completely lose sight of the value they bring.

Which brings us, finally, to tankers. Thinking from a global perspective, tankers are the mega-equivalent of delivery trucks. Tankers get no respect either; they are viewed as grubby things full of oil or rats or both, and that, we have been led to believe, if you watch them long enough will burst open and contaminate the ocean with a giant oil slick.

Like most other elements of the energy delivery system, we should be thankful beyond belief that these things ply the oceans as safely as they do. They distribute many things, but we'll focus on fossil fuels here.

Most petroleum is produced a great distance from where it is consumed. Most of it is produced on other continents. There is no other way to move it than by tanker. The alternative is not to move it at all, and how much fun would that be? As with many things in life, we wouldn't appreciate them until they're not there.

More than half the world's petroleum moves on tankers across oceans or other waterbodies. Tankers are incredibly safe from a leakage perspective. Of course, we only think of one thing when we think of "tankers carrying oil" and that is the Exxon Valdez spill. That was indeed unpleasant, but it was also almost thirty years ago, and in the interval since that leaky day, thousands of tankers have plied the waters safely. Almost no one notices this feat.

Coal is moved across oceans by large ship as well. While slightly different from an oil tanker, a coal ship serves the same purpose, and is just as vital in the supply chain. We won't spend too much time discussing coal ships though because they do not play a significant role in the war on fossil fuels. Burning coal, yes; transporting it, not much.

PIPELINES

For those in the business, the hatred of pipelines has been the most perplexing. It is hard to think of a more innocuous object, and they've been around for centuries. Well maintained pipelines are far and away the safest method of transporting industrial products in huge volumes. It takes a lot of effort to even know where they are.

Pipelines can be problematic under several conditions. You may have heard, through commercials or by seeing signs posted around the countryside, exhortations to "Dial before you dig" or some such. The reason for this is the incredible latticework of pipelines that are under our feet. In one form or another, the following things are moved via pipeline: water, sewage, crude oil, gasoline and gasoline-like products, electrical cables, natural gas, and anything else that is better off underground than being subjected to weather or humans. We need pipelines because moving all this stuff above ground is simply not feasible, or because the dangers to the pipe are too high. Above ground, things happen, by accident and by idiot; it is almost guaranteed. It is far safer to have all these things transported underground.

TRANSPORTING ENERGY – OPTIONS AND LEARNINGS

The biggest risk to underground transport, other than errant shovels or industrial digging equipment, is corrosion. This can happen from inside the pipeline, if what goes through is not carefully understood and monitored, or outside the pipeline, where soil movement can cause stress, or cause rocks to rub against the exterior, or where water can get into coating imperfections and cause rust or other decay. These are operational risks that pipeline companies are expected to be on top of, and generally, they are. There are hundreds of thousands of miles of pipelines under our feet, and they usually work as designed. It's no different in a way from the network of water and sewer pipes running through your house. Are we alarmed by having sewage coursing through a wall within feet of your dining table or within inches of your head? Of course not, and it is a risk we don't ever think about, even after having a few beers and hanging ten works of art in your house by pounding nails straight into the wall and not really knowing what they will encounter. As with pipelines, if designed and operated properly, there is no reason to think they won't function safely.

The importance of pipelines has to do with the magnitude of product being moved. The quantities are so incredibly vast that there really is no other way to move them. In lieu of pipelines, there would be far more activity above ground that would be far more dangerous overall.

What is particularly interesting about pipelines though is how they have taken on a life of their own as some sort of arch-villain, developing a reputation far beyond any actual concrete dangers. They have become so significant they will get more attention in the chapters ahead.

RAIL

If pipelines were ever deemed unacceptable after, say, some big incident, the next most likely transportation method would be rail. Rail systems are designed to haul large industrial quantities of various things, including dangerous chemicals. Their dedicated tracks mean large loads can be optimized without being hindered by road rules or traffic regulations or conditions. Trains can carry hundreds of cars efficiently with just a few locomotives.

As far as petroleum goes, rail is feasible but definitely has drawbacks. Train derailments and accidents are relatively common, and spills can occur depending on the severity of the incident. On the bright side, the spills from rail car incidents tend to be isolated to one or a handful of cars, which limits the size of the environmental contamination. On the dark side, rail lines tend to be constructed along rivers or other waterways, meaning a spill is more likely to get into a water body than from a pipeline incident.

On the really dark side, and I mean really, really dark, petroleum or other flammable loads can have catastrophic consequences. The catastrophes may be infrequent, but they can be gut-wrenchingly horrible, as in the Lac Megantic train wreck in Quebec. In 2013, a seventy-four-car freight train carrying a highly flammable grade of light oil (with an explosive gaseous content) went rogue when a number of braking fail-safes all failed, and the unattended train accelerated down a hill and into a town. It exploded in a vast fireball that killed forty-seven people. It was a tragedy of intergenerational dimensions. As with the water-related incidents documented previously, this one was the result of a chain of small failures that, each in isolation, may not have amounted to much, but in conjunction were catastrophic.

In general, however, rail transport is safe, but is quite ineffective relative to pipelines. In comparison to a large oil pipeline that can carry 300,000 barrels per day, a rail car can carry approximately 700 barrels. To carry the same volume as this sample pipeline, about 430 rail cars would be needed, or something like four miles worth of train. Some pipelines can carry double that, so we can quickly see that moving those quantities by rail, day in and day out, is not practical, especially considering that many other industries want use of the same track.

TRUCK

Trucking petroleum might seem like a reasonable idea, because trucks are everywhere and we're used to them. We also can see that the size of a spill from a truck related incident would be even smaller than a rail car. From the perspective of accident devastation potential then, trucks would appear to be preferential to pipelines or rail cars.

But as with rail cars, the size of the transport relative to the quantity in motion renders this method quite ineffective also. A truck can haul only perhaps 30-40 barrels of oil, so to carry the same volume as our standard pipeline's 300,000 barrels per day would require ten thousand trucks. Imagine trying to pass that procession on the highway.

TRANSPORTING ENERGY – OPTIONS AND LEARNINGS

The small size and relative nimbleness makes trucks useful for hauling smaller loads of petroleum for shorter distances, such as gathering oil from producing wells or for transporting refined fuel to gas stations. As such, they are a vital component of the energy distribution chain, but are almost never exciting and so you probably don't care.

POWER LINES

This one may not have leapt to mind, because we have been preoccupied with anti-fossil fuel messaging. This attention-weighting is ironic in a few ways, the first being that pipelines and fossil-fuel infrastructure are becoming impossible to build, while new major power transmission lines are becoming a growth industry. It will be interesting to see if new power lines encounter the same resistance as pipelines do.

That brings up an even bigger irony with respect to energy infrastructure. We despise pipelines because we *know* they're trouble, we think we do anyway because we hear that relentlessly. But what about power lines?

Let's look at that differently by asking the same question about each. Would you like to live over a major pipeline, or would you prefer to live under a major power line?

You may live quite close to a major pipeline, and could do for decades and not even know it was there. If one were to ever develop a small leak, you still might not even know. A bunch of equipment would show up, there would be an excavation and a lot of big metal shuffling around, and it would be all sewn up again and you would likely have noticed the smell of diesel exhaust more than what was in the pipeline (this is a fairly reasonable scenario for a small pipeline leak, and the majority of pipeline leaks are extremely small).

Would you know if you lived under a power transmission line? Of course you would, and it is stupid to even ask. You would have known that the first time you drove by the property, long before putting an offer on it, and that offer would without doubt be diminished because of the power line. That is if you would even consider living under one, which you wouldn't.

No one likes living under power lines. They make us nervous. No one knows what the impact of spending significant time in a massive electrical field will do to us, and if anyone tells us they know, we don't believe them.

Lunar cycles and the gravitational pull of the moon affect tides, and this phenomenon is the root of the word "lunatic." If the moon can affect water levels and make us go crazy, who's willing to gamble on raising their family in a massive electrical field? What will 24 thousand volts do to Grandpa's pacemaker? Who wants to find out? Does it affect the dog's cognitive faculties? Is a teenager living under a high voltage line like two teenagers?

Yet not only are power lines necessary, they're going to become even more so as we move to renewable energy. Every solar field and wind farm needs to be connected to the grid, a grid which is shaky even now, and is due for massive expansion.

But expanding grids runs into the same **NIMBY** problem as anything else. And it's not just **NIMBY**ism either; some of the impediments to new construction are simply due to the greater attention that is paid to the native habitat (not to mention native as in indigenous habitats). If you're in charge of scouting routes for a new infrastructure project, it is a dark day indeed if you stumble across an eagle's nest in the path. The naturalist in you might rejoice, but not the part of you responsible for a timeline or the paperwork.

FINAL WORD

If we don't want to move energy, we have two choices: move to the source of the energy, or stop using it. It is critically important to understand that limiting supply does not equate to stopping use. That is unequivocal.

20.
The futility of fighting pipelines or infrastructure – energy has to move

STOPPING THE FLOW OF A RIVER WITH A FISHING NET

Imagine you're one of the new breed of west coast entrepreneur celebrating the legalization of pot—not a stereotypical hipster but a more palatable working stiff variety. You spend your days at a relatively meaningless job and the rest of the time high as a kite, with a joint in your mouth and not a care in the world.

One day, you're tending your beloved plants on the windowsill, and a thought strikes you. You think about what you've been reading in the news about how the marijuana industry is now freed from the shackles of the law, the greatest day of your life, and how entire businesses are springing up around the new pot industry. Demand is soaring. You've been downtown and seen the rows of amateur vendors, the ones who start the day perched on stools behind folding tables with cute little handmade signs advertising the day's offerings in front of plastic bowls of inventory. You've seen them finish the day bent right out of shape, dancing with each other in front of their displays, most of the inventory gone and five bucks in the till (I have seen this, and it is very funny). You think, I can do better than that! A little self-control and one could be the undisputed business titan of the block.

So you drop thirty bucks at Walmart for your display and make up your menu board. You show up at the downtown ground-zero of the outdoor markets, set up your table, and you're in business. Now that you're one of the merchants, you notice not just the bozos, but some serious merchants.

Some are professional and well organized; some look like the Russian mob, some look like developing country coup-veterans. Whatever; demand is high and there's room for everyone.

Then one day protesters show up, loads of them, with signs voicing great displeasure at the downsides of pot. Much to your astonishment, they protest your stand only, and none of the others. They're chasing away customers, singling out your product in particular as bad stuff. You're not sure what to do—they're not even hassling the obvious criminal-types. Just you. They show up at your apartment, finding ways onto your balcony to desecrate your plants. Eventually your business is dying, the protesters are celebrating, and you're wondering if you should pack it in and put your table up for sale. More fundamentally, you are utterly baffled as to why anyone would think that wiping you out will help the health of the general population.

That, in a nutshell, is the pickle the energy infrastructure business is in.

In the public arena today, there is an effort to do good that is so sincere and heartfelt that it is almost heartbreaking to point out how futile it can be. It's as though you saw a child trying to revive a dying pet fish with ice cream.

The efforts to which I'm referring are the blockades of various energy infrastructure projects, such as pipelines, on the assumption that this will curtail demand and fossil fuel usage will slow, or stop, and that the planet will be saved from heating up. But the world doesn't work that way.

GET OVER IT: LIMITING SUPPLY DOES NOT STOP CONSUMPTION

We've been through this before in history, this notion that we can solve a social ill by cutting off supply. Here are two of the biggest fiascos of modern times.

Almost a century ago, the US government decided alcohol was a bad idea and proceeded to ban its sale and consumption. That it had been around for thousands of years didn't faze bureaucrats, who possibly were taking advantage of the final days of alcohol to get so drunk they thought they could pull it off. Faster than you can say "underground economy," the illicit trade in alcohol took off, and most adults became criminals upon hoisting their first shot of moonshine. Prohibition, as it was known, was a fiasco, and eventually the government repealed the plan once it realized how silly it was.

THE FUTILITY OF FIGHTING PIPELINES OR INFRASTRUCTURE – ENERGY HAS TO MOVE

The only lasting cultural artifact from the era is NASCAR racing, which was formed by hotshot moonshine runners who not only transported illegal booze, but competed against each other for bragging rights. It is helpful that the gods gave us such a garish reminder of the difficulty involved in cutting off supply of that which is in demand.

Not a bunch of quitters though, those US politicians. Within a few decades, they were back at it, this time turning their full might and fury to the war on drugs, which to date appears to have been a war on their feet, which are now full of bullet holes.

The war on drugs began in 1971, and as a homework assignment, I'll leave it to you to determine how well that's worked out. Torching every poppy field in Afghanistan might feel good and offer great photo ops, but it simply drives up the prices for the particular drug in question, which attracts more growers, and round and round we go.

The world's oldest profession has been harassed similarly and with the same results. Arresting prostitutes is one of the dumbest things police forces have ever done, but they did it for a long time and still do it in some places. Turning the women into criminals does nothing to prevent customer demand, and anyone who thinks that demand can be erased has led a spectacularly sheltered life.

So it goes with fossil fuels. The world's consumption of fossil fuels continues to grow, and the sources of fossil fuels continue to move around. Blockading a pipeline is utterly pointless; the only way it would work is if that pipeline controlled access to the entire resource. Other than pointing out that that would be one hell of a pipeline, there is no point in exploring that avenue further. People may have other reasons to protest pipelines, but strangling supply should not be one of them because it won't work. It never does.

THE WORLD LOOKS A LOT DIFFERENT THAN FIFTY OR A HUNDRED YEARS AGO – A STUPID COMMENT BUT WE FORGET WHAT IT MEANS

It is blisteringly obvious that the world has changed a lot in the past fifty or hundred years. Chalk that up as one of my far-from-insightful history lessons. One of the biggest changes is our view of the changes; we now have instantaneous windows into the entire world and almost anyone can get around the planet in less than forty-eight hours if they can handle the ticket fare and the challenges of modern air travel.

For the purposes of what we're trying to talk about here, it is worth noting the differences in energy flow patterns. The path of energy from production didn't just change direction; it grew by magnitudes of ten or a hundred—and also changed direction.

As with the myriad systems discussions so far, pockets of new high-density populations have sprung up at random places around the globe. The industrial revolution and the rise of North America as an economic powerhouse caused untold shock waves to the world order that had been somewhat stable for the previous several centuries, from an urban perspective—London was London, Mumbai was Mumbai (yes, of course it was Bombay, but don't be so pedantic). Those cities grew like everything else, but they didn't move around much, and existing ones tended to grow more than new ones appearing in the middle of nowhere.

Every single one though became a sponge for fossil fuels. When fossil fuels meant coal, that was one thing. Coal could go anywhere, by donkey if necessary. Oil was somewhat similar, in small enough quantities, but no populated center consumes so little oil anymore, so donkeys are out. Therefore, storage and transportation infrastructure is required, meaning massive tanks, truck loading/offloading facilities, and pipelines.

Natural gas is another matter entirely, because it can move in no other way than by pipeline (Although I did see a picture of a Chinese youth carrying some sort of massive balloon about twice his size down a rural road, reported to be filled with natural gas). Oil can move by pipeline, by rail car, or by truck, but not natural gas.

In reality, there is no safer way to move large quantities of petroleum than by pipeline. We therefore should be thankful for them, because we presently can't live without them, not in the way we want. The argument that we need to get off fossil fuels must be segregated from the fact that at present we consume such ridiculously vast quantities that there is no other way to move them.

Therefore, we find ourselves back at the systems issue again. As populations grow, and more importantly as consumption grows, new infrastructure

THE FUTILITY OF FIGHTING PIPELINES OR INFRASTRUCTURE – ENERGY HAS TO MOVE

is required to keep everyone more or less content. We might wish we were not so dependent on fossil fuels, but as long as we are, we need to build these supply chains to keep everyone warm and fed and to provide the jet fuel for indispensable trips to Cancun.

In sum then, the world simply requires a lot more fossil fuels than it used to, and that means we need to build the means to get it where it needs to be. That is the first half of the infrastructure story.

YOUR ENERGY DOESN'T COME FROM WHERE IT USED TO, AND WE MUST DEAL WITH IT

For someone with an aversion to history, talking about where fossil fuels come from is a bad idea because not much is older. Fortunately, we can skip the prehistoric musings that geologists can go on about in their sleep, and focus on where the produced fossil fuels come from.

Before that, it is necessary to take a quick narrative detour to point out a few things about where fossil fuel comes from, geographically, because that is pertinent to why we need to move it. Coal, we know about; it comes from holes in the earth from which emerge miners coated in black dust like Kentucky Fried Chicken is in batter, and these miners bring forth copious amounts of the flammable black rocks. We also know it can be transported easily, that it doesn't cause messy spills, and that if a coal barge went down in middle of the ocean there would be a black pile on the ocean floor, and that's about it.

Oil and natural gas, on the other hand, are more problematic. The section before covered some of the difficulties required to meet growing demand from more people, larger cities, skyrocketing SUV sales, and more gadgets. There is also another reason for the proliferation of petroleum infrastructure.

Petroleum is not an infinite resource and it is not ubiquitous either. The whole history of petroleum is one of "discoveries," which are then developed and produced, and then once that resource is depleted, the industry moves on to the next one.

Discoveries aren't necessarily down the road from each other; they may well be in a different country or continent or in the middle of the ocean. If new oil fields appeared right next to the old ones, the infrastructure game would be much simpler.

But it's not even close. Consider the biggest oil field developments in North America in the past twenty years: the Canadian oil sands and the Texas shale fields. The oil sands were more or less isolated in northern Alberta, and when production doubled, then tripled, new infrastructure was required to move that oil. Massive new pipeline systems were required to connect to the existing major arteries and refineries.

Texas provides an even weirder example, because its new fields *are* its old fields, and that is *still* a huge infrastructure problem. The huge Permian basin, which is the source of almost all of the world's oil excitement these days, has been producing oil since the 1930s but in relatively small amounts. The west Texas region was therefore covered by a lattice of rickety old pipelines that had worked well enough for decades. Then this past decade, with the dawning of the fracking revolution, Permian production has boomed again by millions of barrels per day. Not only has this new production overwhelmed the existing infrastructure; it has far outgrown the pipeline industry's ability to move oil out of the region. We are therefore seeing a new pipeline construction boom in one of the oldest oil producing regions in the world.

So it goes. The world's voracious appetite for fossil fuels means that some oil fields get exhausted but produce on in a slow death. New ones appear, and they could be anywhere, and if there are enough reserves in the new discoveries then the world will connect to them, somehow. Some of the biggest discoveries in the past decade have been more than a hundred miles offshore. Even donkeys refuse to bring back that oil, or if they do, their danger pay requirements are so steep as to challenge the play's economics.

This phenomenon is not going to change as long as global oil and natural gas demand remains; old fields will be depleted and new ones found. What's more, the demand for new infrastructure may even get worse, because the quantities required grow every year.

Unless we adopt some sort of nomadic lifestyle where we relocate masses of people to where the energy is, then we have two choices: we can either accept that fossil fuels need to move from where they are produced to the rest of the world, or we have to do something serious about consumption.

THE FUTILITY OF FIGHTING PIPELINES OR INFRASTRUCTURE – ENERGY HAS TO MOVE

This is the 8,000-pound gorilla in the room, the one as obvious as gravity and yet the one that paradoxically no one wants to talk about.

THE BRIGHT SIDE OF A DISASTER – A SIMILAR SECOND ONE IS A LOT LESS LIKELY

Let's take a side detour for a second and talk about air travel. Once upon a time, people could get on planes without taking off their shoes, their belts, their phones, removing their nickels, and without the indignity of twirling in a circle with your hands over your head while some grim security agent waves a disturbingly high-tech stick at your genitals. Given what we know about humans and can find on YouTube, there is probably a segment of society that enjoys that, but I don't, and I'm reasonably certain you don't either.

The outcome of these bizarre rituals is that flying on a plane is much safer than it was before September 11, 2001, when a band of profoundly deluded losers concluded that commandeering civilian planes and killing everyone aboard was a good idea, oblivious to the fact that any god that would be impressed by such brutal nonsense is by definition the antonym of a god. The world of air travel changed on that day, and while we are not guaranteed of safety, flying has become much safer because one entire tool kit of the army of fools was taken away. They have others, but this one has been disabled. Flying anywhere is now much safer because of this singular tragedy.

Such is the world of continuous improvement that is brought about, inhumanely as it sounds, by disasters and tragedies. They can be brutal and horrible, but because of that, they get our attention, and we learn from them.

It is of course an immense stretch to compare oil spills to terrorist attacks, even for the most vociferous of environmental groups, but the principle remains that unfortunate incidents make subsequent similar ones that much less likely.

Consider the oil-related fiascos discussed earlier. The Exxon Valdez incident made every oil tanker journey around the world safer. In fact, according to Wikipedia, the Exxon Valdez spill resulted in the International Maritime Organization introducing comprehensive marine pollution prevention rules, which were enacted globally. The US Congress passed the Oil

Pollution Act in 1990, which ensured responsibility for spills and protocols for dealing with them. The US also moved towards standardized double hull oil transportation vessels, which effectively reduce the magnitude of spills. In addition, every crew member of every oil tanker would be aware of the incident, and that could only help with awareness and safety precautions.

The BP GoM explosion and subsequent disaster had similar impacts on regulations, so much so that the US government actually disbanded the Department of the Interior's Minerals Management Service, deeming it to have done a poor job of policing the drilling industry. The dissolved department was replaced with three new ones that separated various functions of the oil and gas business in order to have better oversight. You could argue whether three government departments of anything are better than one, but let's give them the benefit of the doubt, that this structural change resulted in better oversight since each group could focus better.

While no sweeping legislation arose from Enbridge's Michigan creek-side adventure, that doesn't mean there were no consequences. A spotlight was shone globally on corporate emergency response plans, training, equipment levels, and inspection records. This happened not only for Enbridge, but for every pipeline operator in North America. The Michigan spill therefore enhanced the operating procedures of pipelines around the world, and made a subsequent spill that much less likely. The Michigan spill led to a new focus on spill simulation exercises like the one described earlier, and each of these exercises in turn led to new findings that increase safety.

The Enbridge Michigan leak had another side effect as well that is rarely mentioned, because few thought about it. This effect has to do with the impact a spill has on the rest of the industry.

YOU THINK YOU HATE OIL SPILLS? NOT HALF AS MUCH AS THE OTHER PIPELINE COMPANIES

Imagine you're a pipeline company that has big expansion plans. Suppose you've been working with customers for years to negotiate long-term contracts, which have come to agreement on terms, and the lawyers are doing what they do in their intelligent and menacing glory. You're getting bid proposals for the long-lead items from the other side of the world that take six months to construct. You've talked to landowners and have been

THE FUTILITY OF FIGHTING PIPELINES OR INFRASTRUCTURE – ENERGY HAS TO MOVE

successful in negotiating much of the right of way. You can almost smell the cash flow.

Then, right in the middle of this, some competitor has a major incident or dumps a bunch of oil into a river. The media goes into overdrive, social media lights up, and the opponents of fossil fuels gleefully take to the web to celebrate.

What do you think happens to your own pipeline plans after that? They get stupefyingly harder, because there will be much additional pressure to halt new developments. The Enbridge Northern Gateway project saw opposition spike after the Michigan spill, to the point that it was killed off altogether.

No petroleum transportation company wants a competitor to have a spill or an incident of any kind. This concern seems almost poignant, to see such a benevolent and loving relationship in the cut throat world of commerce. The reality of course is that the concern isn't out of brotherly love, it centers on the degradation of the entire industry's ability to build anything.

This isn't like, say, phones, where a Samsung exploding battery means more sales for Apple. An exploding pipeline for one company means it will be infinitely more difficult for every pipeline company to build the next one.

Remember that the next time there is an incident. It might make you sad and angry to see an oil slick on a lake, but rest comfortably knowing that it burns the spill perpetrators' competitors even more.

21.
But it never was about pipelines

It may be surprising to hear considering all the protests that have been going on, but pipelines are in a way nothing but a smokescreen.

ABUSING THE HELP

Many well-meaning people are involved in the protests against pipelines. Some are genuinely concerned about the effects of a pipeline spill, or the risk to underground water aquifers in the event of a leak, or the impact of additional shipments on whale populations.

There is good news about such concerns in isolation, however. These are all concrete and discrete issues that can be discussed and dealt with. A lot of infrastructure has been built in the world, and a lot of ships move through a lot of water, and we have found ways to make these things work. It isn't always perfect for the environment; with most industrial activity, there is always some element of risk. But there is a lot that can be done to mitigate risk to levels that satisfy a reasonable person's expectations.

As examples, consider what can be done about the three listed earlier. Pipeline spills can be minimized by placing any type of reasonable standard we want on the care and maintenance of the hardware. To prevent Enbridge Michigan-type spills, we could require that any pipeline past a certain age, or poor inspection record, or whatever standard we want, we could place severe operating restrictions on any line that fails. We could demand that operators have a minimum level of training.

The risk to underground aquifers can be dealt with safely. Perhaps pipelines have a special installation process in sensitive areas, or have thicker pipe, or different coating.

With respect to whale populations, there are all sorts of solutions, there must be, because certain types of ocean vessel are strongly encouraged. Cruise ships are welcomed with open arms, as are other types of freight haulers. The rest of the world has sufficient capability to move half the world's oil production, some 50 million barrels per day, by tanker, and safely; why can't it be done here? Why can't every oil tanker be demanded to have some sort of escort to open waters?

We can go on through the list of concerns, and a few things will happen. We will see that each is something we can deal with, by working together and demanding best practices. Unfortunately, however, another thing will happen: we begin to smell a rat.

All the sincere people with sincere concerns about pipelines have been recruited to join the overarching cause, and that one is a bit of a Trojan horse. What has happened is that the architects of the climate change battle have co-opted those with genuine pipeline concerns to fight a much wider battle.

FIGHTING SYMPTOMS IS POINTLESS, BUT MAKES ONE HELL OF A GOOD STRATEGY

This situation is important and needs clarifying. Climate change is portrayed as a doomsday scenario for humankind based on a temperature increase that will raise sea levels, change climate patterns, and more. But the climate change battle is, for activists, a catch-all framework for combating a multitude of perceived social ills like sexism, oppression, income inequality, and social justice of many stripes.

How can we know this to be the case? A good place to start, and a pretty fair piece of evidence, is when it comes directly from the mouths of the most ardent climate warriors. Here's a quote from a UN research paper[1]: "The discussion of the impact [of climate change] was initially focused on its physical side, i.e. on the impact of climate change on nature. With time, the social impact received attention, and evidence was presented regarding the relationship between climate change and poverty and livelihood. However, the interlinkages between climate change and within-country inequality have not yet received necessary attention."

What that means is that, for this research group in particular, climate change is far more than a physical problem; it is a social one tied to poverty and inequality. Quotes like this are not hard to find.

Now, consider what the opponents of pipelines are up to, or view their efforts from a ground up approach. Arguments are made that pipelines are unsafe, particularly if carrying certain oil that is deemed "dangerous," which to date has meant oil sands oil (and this is a further paradox: why has oil sands oil has been singled out from all the world's crude grades for specific vitriol?). When these arguments are protested by the petroleum/pipeline companies, the industry points out that there is no particular inherent danger in a pipeline, and that it is the safest method of transportation. So, the argument shifts to the quality of the crude, that some types are abrasive because they may contain sand or corrosive because of some chemical, and when this is shown to be untrue the argument shifts to the effects of a spill, and after that to indigenous rights, and after that to tanker offloading facility risks, and after that to whether single hull tankers are acceptable, and on and on and on. Each argument is a distraction; it is a point at which to engage fossil fuel advocates in distracting debates (which fossil fuel advocates almost always lose because they suck so badly with social media).

For the climate change architects, they aren't serious about any of that stuff. It is all a distraction and a technique to isolate fossil fuel advocates as enemies of the common person, and reinforce the feeling of terror we're all supposed to face at the thought of the planet warming by a certain amount.

Pipeline wars are therefore a sideshow by which useful citizens can be recruited to a cause they might not have signed up for otherwise. When protesters line up to stop a pipeline, it may not be because the pipeline is bad, but because global warming is bad. But of course they don't openly say that. It might seem self-evident, and some protesters would tell you outright that it's a global warming fight, but they would not venture into a discussion about the link between fighting a pipeline and redistributing income. That doesn't sell with the general public, but fear definitely does, and when fleets of scientists point out that we are all doomed if global temperatures rise, then it's mission accomplished.

This matters a great deal if we really want to make true environmental progress. If the straw-man arguments against pipelines dominate, then we get both sides pouring vast resources into pointless debates about whether pipelines are safe or not. That has nothing to do with the fact that oil consumption rises steadily every year, and arguing about it is as relevant as arguing with an armed bank robber about the weather.

If the goal of climate change warriors is to effect societal change, and not just physical change (a shift that the UN report quoted above confirms), then an entirely different discussion is warranted, and pipelines/pipeline spills/ocean spills/fossil fuel consumption are all time-wasting battlefields. Any working group made up of scientists, engineers, government representatives, first nations, and businesses could come up with working plans to mitigate actual concrete pipeline fears, if that were the goal. Social inequality and issues have absolutely nothing to do with the frequency of spills from the world's gargantuan web of pipelines. If environmental contamination were the real target then the vast chemical swamps in China, where rare earth metals are mined to make cell phones (check out Baotou in Inner Mongolia[2] for gruesome pics), which make the petroleum contamination sites look like pots of chicken soup, would be the proper focus of their ire.

Do you attack a big problem like climate change by spending vast quantities of time and resources on the part that matters least? Do you fight drugs by putting three-quarters of the police force at the shopping mall, because it's a known fact some transactions happen there?

Of course you wouldn't, and if the goal was a pure mission to lower greenhouse gases, which are a global problem, one would look no further than the 80/20 rule, which seems to work quite well in a million other spheres.

You are most likely aware but in case not, the 80/20 rule is a guideline for addressing problems that estimates that 80 percent of a solution can be found by addressing 20 percent of the causes. In other words, for many problems, a few members of a population account for the majority of the problem.

What if we were to get rid of the worst 20 percent of power generation facilities? That is, remove the biggest pollution culprits? What would that cost, and what would it get us? These are important questions, if one's motive is to combat climate change. Now, what does spending the same amount on fighting pipelines in North America get you? Nothing, except publicity and a soap box from which to sway the argument away from physical realities towards sociological ones.

As always, this should not be construed as a defense of past practices of the petroleum business, or that pipelines will never leak, or a claim that the oil sands tailings ponds are actually chicken soup. I love chicken soup, and despite having never been there, I can say with certainty that tailings ponds are considerably less attractive.

It is also not intended to be a disservice to honest environmentalists and environmental groups that are working to elicit change. It is not even intended to be a disservice to dishonest ones—they are working towards social agendas and that is their right and prerogative.

But it is necessary to act as a bouncer and wade into the pit of drunks and start dragging out a few of the more pigheaded combatants. There is no intent here to paint only radical environmentalists as being the culprits of disinformation; it's just that the mainstream narrative regarding fossil fuels is defined entirely by these people. The views of the petroleum industry are hard to critique, because there really aren't any.

Has the oil industry been guilty of environmental abuses? Absolutely. Has it run roughshod over animal habitat, forcefully overcome objections of locals, and worked with governments to ensure it gets its way? No doubt that at some point in the past, this has happened. Has it placed profit over safety and good corporate citizenship? At some point, I am sure it has. Has it used behind the scenes influence to benefit itself? I firmly believe so.

At some point though, it is not helpful to bring up historical wrongdoings as a pretext for interfering with the energy systems that the world is hopelessly addicted to today. If oil companies destroyed caribou habitats fifty years ago, but today toe the line with regards to stringent environmental standards regarding those same beasts, there is no value in punishing that segment of industry simply because it used to behave a certain way, and operated under a different and aloof framework.

The tactics that the environmental movement wants to use to fight climate change are up to them; they have their own motivations. There is a desire to have the climate change movement address a host of social ills, and sociologically perhaps some people think that's reasonable to discuss. There is a downside to that though: to discuss social ills and search for cures to them is to open that big red door festooned with warning stickers, danger symbols, radioactivity notices, and that little skeleton symbol for corrosive materials. This is the door labeled Politics.

At this point, it is the right place to make one thing unequivocally clear. If politics become a major ingredient of the mix in vital energy discussions, we are doomed to a nasty, inefficient, and frustrating future.

That is not to say that politics won't be involved; politics pervade everything. We need intelligent policies for pollution and energy saving purposes, but politics can't be *central* to the discussions.

The problem with politics is, well, do I really need to point out what the problem is with politics? Let me put this another way: to move forward in the energy game, we need constructive, progressive ideas that set aside a lot of preconceived notions, and to a certain extent we need to forget about the past. That camp needs people. Are you in or are you out?

To make the debate political will virtually guarantee that nothing will be done. We've seen that movie a million times. If a right-wing party wants to implement something, a left wing one will hate it and fight it, and vice versa. These battles are ridiculously easy to slide into, and it's happened with the fossil fuel debate to all of our detriments.

In many ways, the hysteria created by fighting against pipelines and protesting fossil fuels has set back environmentalism by a decade or two, because it's created good guys and bad guys. No one wants to be the bad guy, so they fight each other instead of the problem. Imagine if all the energy that has been put into posturing and fighting had been spent on energy efficiency.

Time then to get focused. What *exactly* is the problem with fossil fuels?

22.
Habitat destruction vs. contamination vs. pollution vs. global warming

Hopefully at this stage, we can if not completely agree, at least see some common ground with respect to the environment. Seven billion humans have a huge impact; it is difficult to remake all those systems that keep all us humans alive; we must at some point remake these systems, and that we need to take very seriously our impact on the planet. Currently, this has come to mean to be concerned with climate change.

But battling climate change appears to be a devilishly hard task, not just the physicality of it but the cock fights that spring up instantly. Because cock fighting has gone completely out of style, we should veer away from it.

A good place to start is to break down what the actual problems are. Climate change, as it turns out, is a catch-all for a litany of problems, not just environmental degradation. That's not just me saying that, the UN did also, as was seen in the last chapter (and so have many, many others). Starting from the thesis that the environment should be the primary concern, there are a number of ways it is harmed: by habitat destruction, by contamination, by pollution, and finally by global warming. It is important to look at the bigger problem (climate change) in terms of what is causing these individual problems, because as it turns out each of these issues has a different solution.

By way of analogy, here's a little roadmap of the relationship that each destructive force has to humans, or to one specific human anyway:

Habitat destruction – eradicating natural habitat by, for example, mowing down trees to make materials to create something like a diaper

THE END OF FOSSIL FUEL INSANITY

Contamination - what happens to that diaper

Pollution - where that diaper goes when it has served it's revolting purpose

Climate change - how the diaper got created, how it got to Walmart, how we purchased it, how we got the food that went into the top of the little machine that made all that material that filled the diaper, how that little machine turns into a big machine, and the billion things it consumes along the way, times the many billion people who do all this (turn into big consuming machines, not wear diapers).

Concerning the current "us vs. them" environmental wars, we need to do a bit of cataloguing too. What are we really fighting, and what is the best way to do it?

To start the analysis, we need clarity. We need to look at what specifically we're fighting against, and what the best solutions are. Scarce resources used wisely can have ten times the impact, if used with a bit of foresight. Since this book is supposed to be focused on fossil fuels, we'll look at the issues through that lens and try to evaluate where fossil fuels stand with respect to each.

HABITAT DESTRUCTION

The fossil fuel industry has in the past had significant impact on natural habitats in both overt and more subtle ways.

If you want overt examples, google images of old west Texas oil fields, or the Kern River Oil Field in California. These scenes are enough to give the phrase "post-apocalyptic" an ugly new meaning. They are not alone either; many of the older densely developed oilfields of the world have a similar rusted-metallic-boneyard-covered-in-filth look. The ones in the US just happen to be visible. There is little doubt that Russia, to pick an easy example, has examples that would make the US ones look like operating rooms. The Nigerian Delta is another poster child: an oil-encrusted moonscape region of pipelines and makeshift oil-stealing camps that will probably never host the Olympics, unless sports take a weird turn someday.

HABITAT DESTRUCTION VS. CONTAMINATION VS. POLLUTION VS. GLOBAL WARMING

In the middle of the twentieth century, environmental standards weren't what they are now, in the western world at least. In the rest of the world, environmental standards were whatever big oil said they would be. I'll give you a hint: the bar was low.

On the other hand, to compare the environmental standards of the day when those developments took place with the standards of today is like comparing a computer with a chalkboard. In fact, times have changed so much that the animals now rule the roost. Activity in many forested areas now only occurs when the wildlife is good and ready, or when they are done doing things they'd rather do in private like enjoying the breeding season or welcoming the corresponding babies. Wolves don't give them such consideration, just so you know.

Habitat destruction is still a problem, but is slowing as a major one. Not in the local sense; we are still losing some forests and farmland to development, but in the global sense. Most people are flocking to cities around the world, meaning less invasive settlement, and we are acutely more aware of critical wildlife habitats. Don't take this to mean that habitats are not still being destroyed, but we are reaching some sort of status quo. The frenzied global search for petroleum is winding down (and I'm talking in terms of decades here, not years), and we know where most of the world's minerals are. What I'm referring to are huge tracts like Canada's boreal forest, the US Great Plains, the Mongolian Steppes, the Serengeti Plain, Patagonia, or Siberia—huge regions that no one seems to want to live in. Some environmentally critical areas of the world, like Brazil's rain forests, are still under threat and are being cleared, but that is happening against the vocal outcry of much of the world and there is hope it will be slowed or halted. Cities continue to grow, but if global populations can stabilize, there is hope that the bulk of the world's wild places will be able to stay that way. It can be hard to return developed areas back to nature, so we can hope for at best some sort of status quo on this topic. Well, that's not quite true. We can hope for some serious reforestation efforts, and those are a distinct possibility. Some decry reforestation as ineffective because the areas may be logged again, but that seems to be a unhelpful way to look at it. What is better, to not reforest at all?

At any rate, one aspect of fossil fuel development that affects the environment is and has been habitat destruction. This has been a big issue, and is currently much less of one thanks to vastly more environmental scrutiny. We can therefore conclude that this negative aspect of fossil fuel development is not a particularly large component of the global warming threat, from a forward-looking perspective, at least not compared to the brouhaha over transportation or other activities that use a lot of fossil fuels.

CONTAMINATION – THE RESULT OF DOING SOMETHING BADLY

This is a depressing topic and a hard one to fight or eliminate. Contamination comes from glitches in industrial systems, and there are a great many of those. Think of all the industrial processes going on in the world right now. Think of how much petroleum is being shipped through pipelines and trucks and refineries and tank farms. And that's only a fraction of it.

Look around at all the stuff. Whether you're currently sitting on a plane or on the toilet or hanging from a trapeze (hey, I've spent my time on YouTube, some of you are pretty freaking weird), you are surrounded by things that came from manufacturing processes that involved moving a lot of raw materials that are often unpleasant or dangerous. The same is true even if you're camping: how did the aluminum and synthetic fabric of your tent find its way from a deep mine somewhere on earth or a petroleum reservoir a mile below the surface to your nature-loving hands? Sure, the amount of aluminum in your tent frame is, cosmically speaking, extremely tiny and weighs no more than a good mouthful of sushi, but think of even that one tiny element. To get you those few aluminum rods required: exploration activity to find the aluminum deposit, capital to be raised to finance the building of a mine, mine construction, the ore to be extracted, processed, hauled to a smelter, trucks to haul the aluminum to pole manufacturers, and trucks to haul the poles to the stores or you. Every one of those steps involves usage of massive quantities of petroleum and chemicals which need to be found, processed, and moved as well, and each step is rife with the opportunity for contamination of some sort.

I know you're mentally exhausted from that, but if we want to eliminate industrial contamination, *every piece of that chain has to operate perfectly*. Every one. You don't even want to think about how formidable a task that is. You think you're tired from listening to this diatribe? Well, what about the tens of thousands of workers and truckers and product handlers in the chain above who, if they get overtired or overwhelmed, can cause a serious industrial accident or spill? Every instance of a person operating something is susceptible to the failures that come along with human beings. We get tired, stressed, cranky, injured, hungry, and even distracted by mosquitoes or a text, and next thing you know there's a dump truck upside down in a ditch leaking oil into a stream. Industrial accidents can, and do, happen anytime and anywhere, and can have massive environmental consequences faster than you can say, "What do I look like, an amateur?"

That's how hard it is to eliminate contamination. A mighty effort goes into it, but perfection is unfortunately not something we can expect.

So what do we do, throw up our hands? That's one option, unless the industrial accident in question removed them. But there is also a reason for some optimism. As was pointed out earlier, each incident provides a lesson for the entire world, and the odds of the exact same thing happening again are significantly reduced. That's why sometimes, if you ever are involved in building something, or get to work with a tradesman or engineer, you might notice on a piece of equipment or building material some little aspect that seems to be there for no reason, a little flap somewhere, or a guard on a blade, or notch in a component, or even some odd little habit that seems to make no sense. But the worker will explain that that little whatever is to prevent some calamity that you would never have thought of in twenty years, and probably the worker wouldn't have either, except that at some point someone had it happen to them, and the whole world got to learn from it.

At the end of the day, contamination is a problem that we need to steward and continuously improve, but it is not a significant part of the climate change discussion.

POLLUTION - THE SUM EFFECT OF A CONSUMER'S LIFE

Pollution is something else again, and in some ways the most exciting of this environmental shop of horrors. Why would pollution be exciting? Because there is so much we can do about it.

Pollution has a box around it; it is finite and measurable and actionable. We can measure the pollution coming from a car's exhaust pipe, and incentivize the company to do something to reduce it. We can do the same for a factory's smokestack, or the number of Starbucks cups we see in garbage cans.

Polluting the environment, however, is not the same as climate change. There is some overlap: emissions from vehicles, for example, cause pollution and also cause CO_2 emissions, thereby getting on the radar of climate change activists. But CO_2 is not pollution per se; it is a naturally occurring substance required by plants. The same is true for methane, which is also fingered as a major culprit in global warming, worse than CO_2 in some respects. But vast quantities of methane occur naturally. One of the largest sources is wetlands, which cover six percent of the earth's surface. That is an incredibly vast area that also forms one of the most important and dynamic ecosystems. An excess of naturally occurring substances is not pollution, just as too much rainfall does not make water a pollutant. Don't misunderstand: excess CO_2 and methane are the global warming story, but from a consumption perspective, not a pollution perspective.

Pollution can be a bit problematic when its point of origin is not clear, but we can still, in broad strokes, find the culprits. Mega-cities know that internal combustion engines idling forever in slow-moving downtowns produce unacceptable levels of air pollution, and we can do something about that. Electric vehicles are a perfect solution for inner city smog. We can trace contaminants or specific chemicals to certain factories or industries; we can trace McDonalds trash back to McDonalds; we can trace a lot of the flotsam in the ocean back to ship jetsam or to which country it originated (not easily, but it can be done in an 80/20 kind of way).

Our way of life is presently responsible for a great deal of pollution, which we can catalogue and analyze and minimize. But pollution is not central to global warming.

GLOBAL WARMING - AN ATTEMPT TO REDUCE CO2 EMISSIONS

Habitat destruction, contamination, pollution: all problems, but all measureable, and if you're adding to them, there's nowhere to hide. If you're ripping out forests to build a mine or subdivision or just because you're rich and it's

HABITAT DESTRUCTION VS. CONTAMINATION VS. POLLUTION VS. GLOBAL WARMING

on your land and you want to, we can see you. You can't pretend that that bulldozer or chainsaw is doing anything other than what it appears to be doing.

If you're contaminating something, we can find you. We can trace things, and measure them, and clean them up, and hold you accountable.

If you're polluting, we can track the source, because pollution is the introduction of something that shouldn't be there. Even if the pollution is large and multi-sourced, like acid rain was, it is possible to narrow down the culprits using the 80/20 rule.

Global warming though, now that's another beast entirely. Is the problem CO_2 emissions? Or too much methane? If it's CO_2, is it feasible or realistic to actually fix the problem where it needs to be tackled? Empirical observations of the attempts so far would say no, because the main sources are, in a way, sacred cows. Chinese and Indian coal burning account for a major part of the world's output, so why are we wasting a second blocking the hundredth largest pipeline in the world from being constructed, the killing of which will not impact CO_2 levels in the slightest? If methane is the problem, what do we do with the world's wetlands, which cover six percent of the earth and produce voluminous amounts of the stuff? Is it acceptable or sensible or reasonable to pursue things in isolation, industrial processes that provide all the heat for our homes but account for a fraction of the methane that swamps do? This doesn't mean ignore the problem of leaky pipes—by all means, fix them—but no one should pretend that that will have any significance on global levels if all the attention is spent on the portion that has the smallest impact. To do that is to do the opposite of the 80/20 rule, to spend 80 percent of your time on the aspect that creates 20 percent of the problem. That version is not nearly as famous, for good reason.

To recap, slowing global warming, if that is the ultimate goal, needs to be a war on consumption, not on fossil fuels, because we presently can't live without them.

To reduce our usage of fossil fuels on a massive scale in a short time frame means depriving a great many humans of them, which would mean dealing with them the way we do with rats, because seven billion people can't exist without them. Is that the way we want to deal with human-created global warming? If you think so, I'm sorry but this isn't your kind of book, and frankly, I shudder to think what is.

23.
Is Green Energy there yet?

As mentioned earlier, in June 2018, an academic group from a university in Holland published an article about a looming threat to many nations, that the value of their fossil fuel reserves would likely be reduced to zero by 2050. Their premise was built on what they see as a rapid build out of green technology.

In April 2018, Portugal made headlines around the world by producing more than 100 percent of its power requirements from renewable energy.

In the fall of 2017, France shocked the world by announcing plans to ban gasoline powered vehicles by 2040.

These are not one-off green energy headlines; they are part of a tidal wave of upbeat reports about the progress green energy is making. The trend is a bit unusual in that mainstream media normally tends to focus on the negative or the titillating. This fascination with positive environmental updates is a bit out of character. But they also can spot a trend, and the media loves a trend, and having the same viewpoint as the rest of the media outlets gives a certain comfort. It is fine to be wrong if everyone else is too, but it is not acceptable to be wrong on your own. The herd instinct is not the domain of cattle only.

Is the herd right? Is that the end of the story, that we're making such amazing headway that countries can meet all their needs by renewable energy alone, as implied by headlines, and are things now so green that a band of academics can declare the world's fossil fuel reserves to be worthless within thirty years? Or is the green-is-taking-over media train barreling along at such a speed that any indication to the contrary is dealt with like an old Buick that's died on the railway tracks?

THE END OF FOSSIL FUEL INSANITY

Sadly, the latter appears to be the case, and the cheering media herd is headed for a cliff. All the major data points flagged above are gross misrepresentations of reality. The "technology diffusion model" that calculates fossil fuels will be soon worthless appears to confuse the ability to switch from telephone land lines to cell phones with the ability to convert trillions of dollars' worth of petroleum infrastructure to non-existent green alternatives with the same ease. For example, as noted above, Portugal was indeed able to produce enough power to meet its needs, but for a very short time under ideal wind/solar conditions (plus, the country has sizeable hydroelectric energy, which is sort of a distant cousin to true renewable energy but still qualifies for the tag). That ability in no way negates the need for a complete suite of fossil fuel power installations that pick up the slack most of the time. Lastly, the French government said it was their intention to do that, with not a word as to how they would achieve it. Only those below the age of five can be forgiven for taking such political statements at face value.

The truth is that green energy is making great strides, but from a near-zero base, and only, in a sense, by cherry picking. Solar power makes great headlines and impressive peak-power-output numbers, but peak-power-output is not a particularly useful metric without batteries to store it.

For balance, consider the following. A recent news story featured a report from a group of environmental organizations that indicated fossil fuels were dying a quick death. What makes the story significant is not just the content, but who wrote it, and the fact the content sounds rosy but does a ridiculously poor job of hiding the bad news, like hiding a rhinoceros by throwing a blanket over it.

This report, from a trio of environmental titans—Greenpeace, Coalswarm, and the Sierra Club—was called Boom and Bust 2018. The report documented, as the title indicated, how coal as a source of power was on its way out, and how this was a victory of sorts in the war on fossil fuels and climate change. The report's executive summary began: "For the second year in a row, all leading indicators of coal power capacity growth dropped steeply in 2017…"

Note the way the opening sentence makes this sound like good news, that coal power capacity growth dropped steeply in 2017 for the second year in a row. Did you catch the nuance? Coal power capacity *growth* dropped steeply, but coal power capacity actually grew in the year, just not as much as it did before.

We see this phenomenon often with respect to oil usage as well. When global demand growth slows a bit, that always makes big headlines, which usually catalogue such trends as "weakening demand." But the reporting and the implications are completely disingenuous and misleading. The fact that the rate of growth slows does not alter the fact that demand overall is still growing.

Germany, one of the wealthiest nations on earth and a pioneer in wind and solar energy, is actually building coal fired power plants, and where not building them buys power off other coal plants from neighboring countries, which is no different than building them.

In other words, fossil fuels are not in any way "busting" as the Greenpeace et al report suggests. In isolated pockets, fossil fuel reliance may be diminishing. Globally, the opposite is happening.

The same problem exists with the academic study mentioned earlier that indicates that fossil fuel deposits will be soon rendered worthless because of the pace of green technology diffusion. The report treats green technology like any other technology, assuming that it can multiply as fast. A cell phone is not comparable to an electric vehicle in any remote sense beyond the fact that both run on batteries. Adoption of a billion cell phones can only be compared to adoption of a billion EVs by someone who doesn't understand the concepts of electricity, vehicles, batteries, cell phones, vehicle batteries, cell phone batteries, or pretty much any physical reality at all. Or by an academic, who may understand these things perfectly within the world of a model, but rather poorly out in the cold hard world of reality.

As a side note, if this ungraceful lampooning of the academics' "technology diffusion model" appears to be an overly harsh indictment of academics and their ability to forecast, consider these two examples. Between ten and fifteen years ago, house prices in the US rose at a rapid pace, creating a bubble of epic proportions that ultimately led to the financial crisis that nearly brought down the global financial system. If you'll recall, banks were so swept up in this alternate reality that they were giving out so-called ninja mortgages to people with no income, no jobs, and no assets so that they could buy houses. The greatest academic of them all, Alan Greenspan, an economist legendary for his knowledge of every last economic statistic down to the number of vacuum cleaner sales in Cleveland, was watching the whole storm brew up as chairman of the US Federal Reserve, and he saw no problems looming at all. After it all blew apart and the world came crawling out of basements to see what was left, there was Mr. Greenspan scratching his head with nothing but a "well I'll be damned" look on his face. A second example emanated from the financial crisis, but this one a bit more concrete in its stupidity. A gentleman named Richard Bookstaber, a "quant" or math genius hired by Wall Street firms to write trading programs and algorithms to outsmart the market, wrote a book called *A Demon of Our Own Design* that chronicled the dangers in these quantitative models that can blow up systems better than nuclear weapons. Ironically, the book came out in April 2007, just before the financial meltdown from the housing crisis unfolded, and on page 54 of his book, Mr. Bookstaber unwittingly shows exactly why the whole thing happened. In that chapter, he explains to us mortals how clever his profession is in designing trades around securities backed by mortgages, and about the risks they have to watch for in building these models. He points out the largest risk of mortgage backed securities is prepayment risk, then goes on to write the mother of all academic torpedoes: "Mortgages have little in the way of default risk, so if you can get rid of the prepayment risk you have a bond that will feed a huge market." Please reread that sentence and consider its implications. Here was one of the most highly regarded mathematical geniuses on Wall Street, justifying his model that was used to value (incorrectly) tens of billions of junk securities that *directly* caused the collapse of the global economic system. Mr. Bookstaber has a PhD in Economics from MIT. He presumably also has a cereal bowl. He should value the bowl more highly.

Anyway, back to the point. What an interesting mess, you're saying, and why should I remotely care. Well, the two articles mentioned above, the academic one about the "technology diffusion model" and the Greenpeace sponsored one on fossil fuels' imminent demise, both made news headlines around the world. Consider what that means. The world now has a suggestion lodged in its brain, and this is how anchoring theory works, that fossil fuels are doomed and soon, and that a band of academics said so and they have PhDs, it says so right there on the page, so it obviously must be true because how could I ever question a PhD with my embarrassing marks and lowly undergrad degree.

However, back in reality, what is happening on the ground is that traditional power plants on which the grid relies for stability are no longer economic, because great gluts of renewable power generated at exactly the wrong time are pushing power prices so low that it makes no sense to keep them running. We now have in several countries a situation where governments are forcing coal-fired power plants to keep running. That is coal, you'll note, not even clean-burning natural gas.

In other words, the system cannot handle what is being shoved down its throat, and that is right now. If the academics referenced above saw even a part of their "technology diffusion model" come true, there would be a systematic collapse of any country that achieved the goal. That's a hell of a way to run a railroad.

Such rapid change is impossible given how reliant we are on ingrained systems. To move forward, we need to understand a little better where we're really at on the continuum of converting to green energy. Make no mistake about it: listen to the guy from the earlier chapter who worked on the electrical grid for forty years over a band of academics whose physical contributions to the economy consist of wearing out coffee cups.

GOING ALL GREEN – UNDERSTANDING WHERE WE'RE AT

We will need two things to go all green, or even get close. First, we will need enough green power to meet our needs, either from large scale batteries or directly from the source (which will mean reorienting much of our usage to the time frames when it is generated, i.e., the middle of the day where solar power is predominant, etc.). Second, we will need to adapt our systems enough to accommodate this. These are both tough topics to discuss meaningfully in a subsection of a chapter; books could be written about either.

In the preceding paragraph are two possibilities for going to renewables on a large scale. If you're paying attention, you may have noticed I committed as grievous an error as the academics did earlier. What you saw right there was the sort of catastrophically dumb analysis that the energy world attracts. What makes it dumb? Because it's a typical oversimplification: we are incapable of understanding the magnitude of the issue. My points may be valid, in an academic sense, but don't really mean anything, in the real world. It is unrealistic to think of moving power usage to correspond with the times of day when renewables crank it out, and large-scale batteries are still mythical beasts.

Actually, this problem isn't confined to energy, and there is another example that illustrates magnificently what we're up against when we toss out oversimplifications.

Currently, over two hundred thousand people have signed up to be astronauts to Mars. A handful of those (at most) would have even the slightest idea of what they're actually signing up for. The vast majority must be either thinking, "Hmm, I'm suicidal anyway, and what a cool way to go," or they wouldn't be thinking at all. They would be like tourists showing up in the Arctic for a two-week holiday armed with only a windbreaker and a selfie stick.

In other words, going to Mars means signing up for a mission with a hilariously high number of ways to die. You must land on a planet with no oxygen and an average temperature of -55 degrees Celsius, or, if you're from the US and starting to pack, -67 degrees Fahrenheit. Your nearest food is 34 million miles away, and if you forgot toothpaste, you will have to wait for the next grocery run, which will be never.

People are signing up for a life-ending experiment that would be fraught with unimaginable challenges, zero margin for error, and if they actually succeeded would wind up in some little pod eating some sort of synthetic protein powder and drinking whatever they could recycle from their own body fluids and let's not even talk about a shower. "Enthusiasm has been growing...apparently they're okay with living out the rest of their lives on Mars," said Mars One CEO Bas Lansdorp. Yes, you idiot, because they haven't the foggiest idea what that would mean. They have some idealized vision in their head that is so lamentably juvenile that it makes one weep for the human race. We idealize and eagerly anticipate our upcoming holiday to Florida, and come back muttering about everything that went wrong, and that's to one of the places on earth geared almost entirely to maximize tourist appeal. I would take a Mars astronaut seriously only if they filled

IS GREEN ENERGY THERE YET?

a backpack and lived on Antarctica for six months with nothing but what they could carry. Outside of that, they're all lunatics.

We get the exact same phenomenon when we say things like, "All we have to do to switch to renewables is move our peak power usage to mid-day," like I did, or that we need a battery system to harness wind/solar power. We have absolutely no clue as to the magnitude of the problem or how difficult it would be to even accomplish five percent of one of these missions, but we can get 200 thousand people to sign up for a one-way ticket into an oxygen-less vacuum with nowhere to go to the bathroom for the rest of their life. So it is incredibly easy to get a few multiples of that number to think, "Yeah, a few batteries, that's all we need."

Perhaps asking if green energy is "here" yet or not is asking the wrong question. It is "here" in the sense that we have a lot of windmills and solar farms that can generate a lot of power, and make no mistake, it's a *lot* of power.

The question should perhaps be, now that we have this resource, what do we do with it?

24.
A way forward, for those who want it

If the debate remains in its current state, with activists on the offensive against fossil fuels and the industry reacting to those campaigns, we will get nowhere. Fossil fuel usage will continue to grow, the campaign of fear will lose its power, and the status quo will prevail until we exhaust cheap fossil fuel energy supplies.

There is a choice though. There are ways forward to prepare humanity for a day when we can effectively reduce reliance on fossil fuels. But we can't get there by infighting; true progress will take all the energy we can muster from both sides. We can't get there by building new green power solutions without considering how they fit in the system. We can't get there if we fail to plan for the changeover, meaning that all industries have to start acting as though the change is coming.

There is no grand simple solution, like banning internal combustion engines or halting all global fossil fuel development. Those are like banning prostitution. But there are many smaller initiatives that will pave the way for a feasible renewable energy future.

To make any headway, we're going to have to make a few assumptions. First, we need to drop the idea that any mass-conversion to green energy is going to happen quickly. If that failure causes the climate to boil, so be it; realistically, there is nothing we can do in the short term. If some global dictator could enforce such a change, we would simply be swapping the potential for a healthier planet fifty years down the road for the certainty of a deadlier one now.

Second, we can't afford the status quo; we do need to embrace new ideas and technologies and not fight them. We need to stop quibbling about the shape of the curve of global warming and whether it was hotter or colder eight thousand years ago. You may be right, maybe it was. Maybe all that CO2 is good for plants. Don't know, don't care. Fundamentally, we can't keep relying as heavily on fossil fuels as we do, because at some point a growing population that relies exclusively on a resource that grows ever-scarcer is going to create a very big problem. It is foolhardy to be caught up in climate interpretive-data-dances when we should be focusing on this fine basket of ideas you're about to read, a basket that didn't originate with me but if you tell people it did I won't be angry.

BRING ON THE FIVE DOLLAR DRINKING STRAWS

I'm not a fast food junkie, but I occasionally succumb to the little pig in my soul that squeals so loud it can't be ignored. It can be outsmarted though, and on a recent pit stop to feed it at an establishment I simply couldn't walk by, I ordered a small salad and chili to go. As I walked away, my inner pig wailing in despair at the thought of upcoming salad, I noticed the counterintuitive heft of the bag containing the culinary treasure. Back at my desk, I cleared a spot, rolled up my sleeves, and unpacked the meal. Simple as it was, the bag's contents included: four small cellophane packages like airplane snacks come in for crackers and salad stuff, a plastic container for the salad, a plastic lid for the salad, a container for the chili, a lid for the chili, a plastic fork in a plastic sleeve, a plastic spoon in same, a paper bag that the chili was in, and a large grocery size bag that everything was in. The drink came with the usual waxy cup, plastic lid, and plastic straw.

Seventeen pieces of garbage for that scrawny little lunch. Some is likely recyclable, but most won't be, because even with standard garbage bins replaced by the multi-option recycling things, virtually no one stands there long enough to contemplate properly whether the waxy cup is recyclable, or whether the paper-but-slightly-less-waxy chili container is organic, or looking on the straw to see if it has a recycle logo. Occasionally you'll see someone valiantly attempt to do the right thing, but after ten minutes and with a lineup behind them they invariably roll the whole damned mess into a ball and shove it into the nearest receptacle.

There is a valuable lesson here amongst seventeen pieces of garbage, pieces that will be a problem for some spot on this earth for infinitely longer than I benefitted from the food it accompanied. The lesson is, put that sorry episode into a global framework. Take that example, times several billion, times the number of times this happens in a week. Think of the costs and materials involved in creating all that junk. Think of how much petroleum needs to be processed to create the plastic for that mountain of garbage that's generated weekly, and think of the petroleum needed to haul it away to a landfill or recycling depot for the few scraps that go that route. Think about the fact that there is no penalty whatsoever for all of this, except the most important: the cost to the manufacturer. It is clearly worth it for fast food companies, and whomever else ships you something like a new piece of furniture that has two acres of cardboard in the packaging, to continue these practices.

If the world's objective is to reduce usage of fossil fuels and help the world's pollution problem, which of the three alternatives on the table make the most sense: block pipelines and other energy infrastructure, convince people to stop flying and consuming, or put an end to this packaging nonsense?

Getting intelligent about these supply chain abominations is feasible, possible, and not that hard, and would have enormous benefits. We just need to do it.

Sorry, foes of capitalism, but the price mechanism really does work. A group of protesters could stand at the airport for fifteen years begging people not to fly with no effect at all, while a hike in airfares can reduce demand and cause cancelled flights. If enough people decide it's too much, we would have a huge environmental benefit.

The same goes for the garbage we consume every day. We need to make it costly for both providers and users of this crap so that they pay attention. I have a nifty Starbucks cup, a reusable plastic one I bought one jubilant day when overcome by environmental awareness. For a good week, I toted it around, getting it refilled with a pleasant little jolt of self-satisfaction. Hey, look at me stepping up. The first visit home to the dishwasher though and the confounded thing is in the cupboard and I never remember it. Why? Because the discount offered for its use is ten cents. I apparently have given up wondering what I would do with that extra dime. But what if a disposable cup like the ones I desecrate the earth with regularly costs, say, five bucks, *in addition* to the coffee? I am reasonably certain I'd be lugging around my reusable cup from here to eternity, and so would tens of millions of people every day, saving tens of millions of cups.

What if a drinking straw cost as much as the drink itself? How many do you think would end up in landfills? What if those drinking straws were biodegradable? As a matter of fact, why aren't they biodegradable right now? They don't need to last for three years, more like three minutes. By the time you read this, the tide may well have turned against plastic disposable drinking straws; a friend recently was offered one made of pasta. How brilliant is that? You could even drink boiling hot liquids through it and cook dinner while you enjoy your beverage. Perhaps I'll leave the creativity to others, but you get the point.

There are countless inefficiencies that can be shaped by proper incentivizing, if done correctly. If done incorrectly, we wind up with overflowing landfills and oceans so filled with garbage we can almost walk across them. But things don't have to be that way.

Strategies along these lines have multiple benefits. They reduce pollution, and they also provide help for those primarily concerned about climate change. Climate change is caused by the sum of human activity, and if an industry like, say, disposable drinking straws could be eliminated, then whole chains of industrial activity would be eliminated also. Few plastics would be created/used/transported, less garbage would need to be hauled, and a lot less fossil fuels would be burned.

HYDROGEN

At this point in the world of renewables, it appears that electricity has won the race, with respect to the future of personal travel anyway. How do we know this? Because the news flow tells us every single day. We hear about "skyrocketing" EV sales, and about record power requirements being met in various countries by solar and wind which feed into the electrical grid.

As with many elements of this whole debate though, we need to be careful about what we think we know. For example, EV sales are indeed skyrocketing, just as sales of top-of-the-line Ferraris skyrocket when they go from three to four per year. In other words, the statistics overstate the importance in the overall market, because EV sales start from such a low base.

A WAY FORWARD, FOR THOSE WHO WANT IT

What does this have to do with hydrogen? It is simply a reminder not to write it off, because it has incredible environmental potential. It is probably the smartest choice as a hybrid fuel in the transition to green energy.

Hydrogen has a number of advantages. Let's get to the coolest one first. A hydrogen powered vehicle produces a single emission: fresh water. Imagine sitting stuck in traffic on a hot day and instead of being nauseated by gasoline exhaust fumes, all you hear is water trickling out of exhaust pipes. I suppose if the traffic jam is bad enough that would be a bad thing from a "need to pee" perspective, but that's an improvement from an "I need to pee and I feel sick" perspective, right?

But I can hear the protests from here: EVs have no emissions at all and are perfectly neutral from a bladder perspective. First, it's not true, because electrical power that charges EVs has an emissions component, but that's not even the pertinent issue. The problem with EVs is of a stupendously larger scale—and here we go back to the systems problem again. As was discussed in an earlier chapter, mass conversion to EVs will stress the existing electrical system enormously, and many components of the electricity delivery chain simply couldn't handle it without a complete overhaul. Complete overhauls of systems that size have multi-trillion-dollar price tags. In addition, the problem remains that renewable energy produces its largest benefit at the wrong time, unless we all could plug in our vehicles at high noon or when the wind picked up. That scheme would go over as well as getting people to swap cell phones for land lines.

Hydrogen would have a similar system overhaul requirement, but the fuel has one massive advantage, one which trumps the whole debate: hydrogen *is the battery*. This relevance of this characteristic is not to be understated, because it is the only scheme that can successfully marry the preferred renewable energy sources with the existing fueling infrastructure. This is because hydrogen can be created at any time by any power source, such as a wind turbine or solar installation. Creating hydrogen is a store of energy, which can then be used elsewhere at other times.

The beauty of hydrogen as a fuel is that it could, in large part, use the existing refueling system. The gas stations and pipeline systems won't go to waste, as they would in an EV switch. Yes, they will need modification;

hydrogen is not as easy to pipe as existing fuels and requires expensive stainless-steel pipe. But it can be done.

If you aren't buzzing with excitement yet, well you better sit down because you will be in twenty seconds. What makes hydrogen such a spectacular choice as a transition fuel is that it can be created by natural gas as well. Thus, the entire natural gas distribution system can be used as the base for a hydrogen energy system, with wind/solar power sources tapped in wherever they exist. How f___king beautiful is that? On top of that, any methane source like landfills that creates natural gas can be tied in to the system as well, further enhancing the green street cred.

No renewables system can function without the backstopping of fossil fuels to meet peak loads, or kick in to meet demand (all demand, remember) if the wind isn't blowing or it's dark out. Fossil fuels are absolutely necessary, and if we're going to use fossil fuels regardless, shouldn't it be the cleanest?

Critics point to hydrogen as impractical and ridiculously expensive to implement. I read one commentary in an oil industry magazine that said the "technology was not feasible," which sounds like what you'd expect from a guy writing for an oil industry magazine. It's not true; the technology definitely exists, the question is only with cost and scale. But as we've been over already, any massive system change is going to require staggering amounts of money. Upgrading the electrical grid to handle wind and solar will cost a staggering amount of money. Running out of cheap fossil fuels will require a staggering amount of money. Feeding and heating 7 billion people will require... you get it by now.

A hydrogen/natural gas system would be cost competitive with any other large-scale energy rewiring project, and would offer synergies that no other could match.

SWITCH FROM COAL TO NATURAL GAS

A lump of coal in my stocking. I don't know where that phrase originated, but it's become sort of a heart-warming threat over the years, a nostalgic but not-veiled reference to what one could expect for a present at Christmas time if you have been behaving badly. It was also an historical reference to coal's ubiquity and utilitarianism; today kids might be thrilled to get

A WAY FORWARD, FOR THOSE WHO WANT IT

this amazing organic art tool for sketching. Dumb millennials. Can't even insult them.

But maybe there should be a concerted effort to relegate coal to, and only to, the world of art and find a new metaphorical insult to hurl at underwhelming brats such as the promise of a stocking full of pencils or two-generation-old electronic devices.

By replacing coal with natural gas, in one fell swoop a significant chunk of the world's excess CO_2 emissions could be solved. Of course, that fell swoop wouldn't be easy, but almost nothing that makes a material global change will be remotely close to easy. Nothing says that this all has to happen overnight either; there is a pretty phenomenal difference in cost between doing something right now and doing something over a decade.

Switching to natural gas is a strategy that should put a similar grin on the face of every global citizen concerned about global warming and climate change. It won't, because natural gas is still a fossil fuel, and therefore not on everyone's "approved" list. That is not just a shame, it is illogical, and is one debate that should not be avoided.

Burning natural gas instead of coal results in 50-60 percent less carbon dioxide emitted into the atmosphere, according to a group called the Union of Concerned Scientists.[1] This union is ferociously opposed to fossil fuels, yet even with their hyper-intelligent brows furrowed with concern, they can't evade the fact that natural gas is that much better for the environment. On top of that, it appears that coal contributes approximately 25 percent of global greenhouse gases. Thus, if natural gas replaced coal usage in its entirety, GHG emissions levels would fall by about 12-15 percent, and the world would survive another day.

CARBON CAPTURE, ONE WAY OR ANOTHER

If the average layperson claims to have come up with a miracle gadget of some kind that can revolutionize transportation/climate change/perpetual motion, the world doesn't pay much attention. The web is full of these people with a revolutionary idea that just needs financing and the world will be saved. You can easily find the home-made videos of their planet-saving ideas and devices online, and each video will have six views and end in either an explosion or a perplexed-looking inventor or both.

THE END OF FOSSIL FUEL INSANITY

But what if I told you about a new carbon capture idea that has actually been constructed and tested and that the company that developed it is partially owned by Bill Gates and Murray Edwards? Yes, that Bill Gates. You may not know Murray Edwards quite as well. He is a founder of one of the largest independent oil companies in the world (Canadian Natural Resources), is a business genius, and his net worth has more zeros after the one than I have teeth.

Both these gents are, to put it mildly, razor sharp businessmen, and both obviously can smell opportunity like a bear smells honey. Sprinkled on top of that pile of investment capital and business acumen is the founder of the company, a Harvard physics professor named David Keith (yes, an academic, but holy crap, he's actually building something). That trio alone, never mind all the other brainpower they've assembled, makes for a formidable team, so formidable that there is not a single idea that has ever passed through my head that I'd be comfortable presenting to them as a good business opportunity.

The company they own, Carbon Engineering of Squamish, BC, has developed a process to remove carbon dioxide from the air and convert it to clean-burning synthetic fuel. These technologies have been developed elsewhere, but Carbon Engineering's possible breakthrough is the ability to do this on a large scale. Their prototype plant has been extracting CO_2 for about a year, and development is hopefully around the corner.

There are more options for absorbing CO_2 as well, including some like mass reforestation, which is about as low tech as you can get. Grab a shovel and a bucket of seedlings and get to work, a process which can often be put into motion without the help of a single Harvard professor. Planting trees can be a highly effective form of carbon capture, particularly since the optimal species of tree can be planted, and the technology is so green and wholesome that it is inconceivable that anyone would complain about reforestation. Well, I suppose plains buffalo might take exception, but that would be it, and a bilateral negotiation would have a high chance of success as plains buffalo seem reasonable.

The point is, our way of life emits vast quantities of carbon dioxide, and there isn't much we can do about that over the next few decades without getting rid of a lot of people. If carbon dioxide truly is the problem, or rather, if it is defined as the problem to the extent that the world deems it problematic enough to deal with, then carbon capture may be the smartest way to go. Just as governments absolutely need the global economic engine to keep turning to avoid defaulting on massive global debt loads, our way of life requires it to keep running on cheap energy. That global economic engine can afford many things, including the cost of developing carbon capture systems, and particularly so if the carbon capture systems generate clean-burning synthetic fuel or green forests.

CARBON/CONSUMPTION TAXES OR REBATE SCHEMES

Oh no, the dreaded T word. Who likes taxes, other than sadists and bureaucrats? Not even dentists like them, and they are as close to kindred spirits with sadists as you can get.

On the other hand, if done wisely (a word that doesn't often go with taxes, but we can have hope), a consumption tax is an effective modifier of behavior. The tax could be similar to the drinking-straw deposit scheme, where the object being singled out need not be fossil fuels but could be anything wasteful.

The trick and the task of governments should be to sniff out these opportunities, to find avenues that are problematic to the environment and deal with them. Why torment everyone with a carbon tax, which hits productive industries, when we could have a program that has citizens pursuing it with a vengeance? Don't laugh, it exists in real life. The bottle/can refund program is a poster child. Why is it not applied in dozens of ways? It works nearly perfectly, and I see it in action daily. An army of street people scours the streets, alleys, and garbage cans every day, looking for returnable cans and bottles. When you think about it, that is an unbelievable accomplishment on multiple fronts. The process is incredibly efficient: those people will lift a dump truck to get a beer can, and they are often seen upside down, feet flailing in the air as they unearth a prize at the bottom of a trash can (I never said it was pretty, I said it was efficient). People who won't or don't work have an opportunity to make a bit of money (and we know they can work because it is work to gather returnable cans) on their own terms, which isn't exactly the sort of career that high school guidance counsellors steer you towards, but it is nevertheless better than sitting on the curb if that is your alternative. I'm not trying to glorify or belittle the bottle picking habit, the point is that the tactic utterly and completely removes from the landscape what would otherwise be trash. It is a technique that could be used over and over. The key is incentivizing properly and effectively. What if plastic, any plastic, had a refundable deposit tacked onto its purchase price? Maybe we'd see squadrons of little scavenging boats plying the oceans, picking up booty just like picking up beer cans.

Inevitably ideas like this often run into walls of opposition or, if not opposition, complaining. It will be too hard to orchestrate internationally. My country didn't cause it, yours did. And on and on. We are being asked to rewire the world's energy delivery systems in order to prevent climate change, and we're going to have to rewire the world's energy systems to deal with petroleum as a scarce commodity. It's all going to be hard. It should be a lot easier to implement GIPRS, the Great International Plastics Recovery Scheme (someone may possibly think of a better name, but I doubt it), than to implement a meaningful global climate change scheme, as the Paris Climate accord of 2015 has shown.

USE WHAT WE HAVE – BUILD A THOUSAND GREENHOUSES WITH OUR WASTE ENERGY

A few chapters back we stopped in to visit Wes and his potato patch, and pondered why Vermilion Energy's waste-heat tomato greenhouses weren't being duplicated elsewhere. The potential to pursue similar opportunities—utilizing resources either going to waste or massively underutilized—is staggering.

Some cities are enjoying urban rejuvenation where volunteer groups are creating small gardens in any available space. The practice is quite amazing when viewed up close: an area the size of a living room can produce a surprising number of herbs and vegetables. Unfortunately, if the mini-garden is too inner-city it also provides stomping grounds for vagrants, who may feel rejuvenated and more mainstream by rolling around in a patch of rosemary, but even then, there is a net benefit to society.

If we think of what can be done on those tiny parcels of land, what about the larger ones? How many CO_2-sucking trees could be planted in some of the larger spaces, such as around traffic interchanges?

A WAY FORWARD, FOR THOSE WHO WANT IT

What if we think bigger, about truly utilizing waste heat like Vermilion is, but all over the place? What if big buildings did the same, or factories, or anywhere with waste heat?

This is where we need to think outside the boundaries of the fossil fuels and climate change debate towards the efficient use of resources. If we used waste heat from industry, wherever available, we could displace, for example, vegetable imports from foreign countries. Think of the spin off benefits of that. We would keep money at home instead of sending it abroad. We would create jobs and new income streams for governments and businesses. We would have fresher vegetables as opposed to ones that have travelled several thousand miles. Speaking of travel, we would save the fuel and maintenance costs on hundreds or thousands of truckloads of stuff coming from, say, California, which would help with greenhouse gases as well as leaving a smaller environmental footprint overall.

If we were incented properly, there are literally thousands of these opportunities surrounding us that would help on many, many fronts.

We also need to stop defaulting the way we do now. When the fossil fuel industry looks at every pronouncement of an environmental group as harmful propaganda, that stops progress dead in its tracks. When environmental groups envelop actual energy issues with social policy initiatives, that stops progress dead in its tracks. Attitude, sometimes, is everything.

Epilogue

I happened to catch something in the news that brought more tears to my eyes than a bucket of onions. In San Francisco, a Judge William Alsup was presiding over a case brought by New York City against some members of big oil for causing climate change. The judge is apparently a bit of a geek and a ham, preparing for a part of the trial by watching National Geographic shows and wearing a "science tie" to trial with little planets on it. He then asked both sides an important question: to weigh the large benefits that have flowed from the use of fossil fuels against the possibility that these fuels may be causing global warming.

The judge's question has that beautiful eloquence one sometimes comes across in legal decisions, comments that cut to the essence of mountains of evidence and raging opinions. What indeed has been the benefit to humanity of using fossil fuels, and now that we are so hooked, what do we do if we want to get off? Conversely, if fossil fuels are the problem, what is a viable solution to the problem that doesn't involve shouting, chaining oneself to a valve, or making up stories to scare people? The best part of the judge's question was that he asked the same thing of both sides: to comment on the benefits and the drawbacks of fossil fuels. Beautiful. However, after the preceding was written, the judge delivered a quick verdict and dismissed the lawsuit, stating that this was a societal question for which every government bears responsibility.

This book is an attempt to encourage that sort of discussion, to paint a picture of how we got here for the purpose of disarming and de-emotionalizing the debate. We don't need a catfight or even a tomato fight—we need solutions. The solutions are there. We need people to focus on making it all happen instead of infighting.

THE END OF FOSSIL FUEL INSANITY

I once heard of a situation that happened between two people having a raging debate on the internet, where one of them changed their mind. I'm pretty sure it's just an urban legend. It is theoretically possible but I don't believe it could be recreated in captivity.

Let's not let this debate become stalemated by entrenched positions. Let's not look at the environment/climate thing as a war on fossil fuels. To start down that path is to instantly guarantee that nothing of substance will be accomplished, or that anything that is accomplished will be ten times harder than it should have been. Fossil fuels are presently our life, and we need to accept that and yet get to work on the replacement. That will take all the available mental energy and then some, and we really don't have the luxury of stooping to try to win an argument.

For some it is too late. They have staked their ground in one of the camps, and human pride will never allow those groups to meet in the middle. The vast majority, the legendary silent majority, will be the ones that bring about change. The world is too complicated to have deal with intractable wingnuts. When you get involved with them on any level, you then have two humungous problems to solve instead of one.

To wrap up, let's go back to Rumi from Sufi again. Or was that Sufi from Rumi? Anyway, about that field. Beyond producing fossil fuels and consuming fossil fuels, between protecting the environment and protecting the energy systems we use, between all these things, there is indeed a field, and as many of us as possible should try to meet there. Thanks for reading.

References

Chapter 1

1. Tabuchi, H. (2017, July 1) As Beijing Joins Climate Fight, Chinese Companies Build Coal Plants, The New York Times https://www.nytimes.com/2017/07/01/climate/china-energy-companies-coal-plants-climate-change.html (Accessed September 23, 2018)

2. Dziedzic, S. (2017, September 5) Electricity market struggling as coal-fired power stations shut down, regulator says, ABC News http://www.abc.net.au/news/2017-09-06/electricity-markets-struggling-as-coal-shuts-down-aemo-says/8875874 (Accessed September 23, 2018)

3. U.S. Energy Information Administration (March 2016) Trends in U.S. Oil and Natural Gas Upstream Costs https://www.eia.gov/analysis/studies/drilling/pdf/upstream.pdf (Accessed September 23, 2018)

4. Cunningham, N. (2016, September 14) World's Most Expensive Oil Project Could Finally Come Online, Oilprice.com https://oilprice.com/Energy/Energy-General/Worlds-Most-Expensive-Oil-Project-Could-Finally-Come-Online.html (Accessed September 23, 2018)

Chapter 2

1. Wikipedia (June 2018) Haussmann's renovation of Paris https://en.wikipedia.org/wiki/Haussmann%27s_renovation_of_Paris (Accessed September 23, 2018)

2. Elsner, D. (1986, December 2) GM BUYING OUT BALKY BILLIONAIRE, Chicago Tribune http://articles.chicagotribune.com/1986-12-02/news/8603310009_1_gm-stock-computer-systems-company-perot-state-of-the-art-auto-plant (Accessed September 23, 2018)

3. (2017, May 15) Petrol cars will vanish in 8 years, says US report from Stanford Economist, Financial Review https://www.afr.com/business/energy/oil/petrol-cars-will-vanish-in-8-years-says-us-report-from-stanford-economist-20170514-gw4r0u (Accessed September 23, 2018)

Chapter 3

1. Casselbury, K. (2018, April 25) What Fresh Fruit Do Americans Eat the Most? SFGate http://healthyeating.sfgate.com/fresh-fruit-americans-eat-most-9593.html (Accessed September 23, 2018)

2. Wikipedia (2018, September 13) Banana https://en.wikipedia.org/wiki/Banana (Accessed September 23, 2018)

Chapter 5

1. Lonely Planet Writer (2017, August) Open road adventures: six epic drives of the world, Lonely Planet https://www.lonelyplanet.com/travel-tips-and-articles/open-road-adventures-six-epic-drives-of-the-world/40625c8c-8a11-5710-a052-1479d2769822 (Accessed September 23, 2018)

2. Office of Energy Efficiency & Renewable Energy (2017, January 30) Fact #962: January 30, 2017 Vehicles per Capita: Other Regions/Countries Compared to the United States https://www.energy.gov/eere/vehicles/fact-962-january-30-2017-vehicles-capita-other-regionscountries-compared-united-states (Accessed September 23, 2018)

3. Sheehan, S. (2018, April 25) Volvo electric cars to make up 50% of brand's sales by 2025, Autocar.co.uk https://www.autocar.co.uk/car-news/motor-shows/volvo-electric-cars-make-50-brand%E2%80%99s-sales-2025 (Accessed September 23, 2018)

4. Mercure, J.-F. et al (2018 July) Macroeconomic impact of stranded fossil fuel assets, Nature Climate Change, Macmillan Publishers Limited https://www.nature.com/articles/s41558-018-0182-1.epdf?referrer_access_token=DPQ9nE6dGXY8k_WL0GzV5tRgN0jAjWel9jnR3ZoTv0NBDAdbQ1RWHSa6L720gc7lUR_z1wT-njPIOyV5lXvFMVIyNMlKx4fgOStd2gybbUXpfV764_dz205QjpB4t-BquTRXKIQ8mR_xyGe95EM1tNvSwwTOUkuXRQw4zO84NVQliyxvu7bPS-vZvOOIlC3TLMlbupeWtl_D9HrZzGiwB84g-a8rSjWL67Ek5WPcG1FtU-cGx_W8zL320pSUemzutUmgkYaBDSH1htKrP2RH63ltRswYRiFW3MazAKhVr_NnXE-kLQymKYjfevLAcSzTtzs5E&tracking_referrer=www.theguardian.com (Accessed September 23, 2018)

5. Elliot, J.K. (2018, June 5) A 'carbon bubble is coming, and Canada's oilsands are doomed: study, Global News Calgary https://globalnews.ca/news/4253559/canada-alberta-oilsands-financial-crisis-carbon-bubble/ (Accessed September 23, 2018)

6. Rowell, A. (2018, March 13) Schwarzenegger to Sue Big Oil for 'Murder", EcoWatch https://www.ecowatch.com/schwarzenegger-oil-companies-murder-2546891443.html (Accessed September 23, 2018)

7. Rosling, H. (2018) Factfulness, New York, FlatIron Books

REFERENCES

8. Watts, J (2017, October 30) Global atmospheric CO2 levels hit record high, The Guardian https://www.theguardian.com/environment/2017/oct/30/global-atmospheric-co2-levels-hit-record-high (Accessed September 23, 2018)

9. Varadhan, S. and Dasgupta, N. (2017, July 4) Exclusive: Indian utility bets $10 billion on coal power despite surplus, green concerns, Reuters https://www.reuters.com/article/us-india-power-ntpc-exclusive-idUSKBN19P1NC (Accessed September 23, 2018)

Chapter 6

1. Wikipedia (2018, March 17) IEC 61850, https://en.wikipedia.org/wiki/IEC_61850 (Accessed September 23, 2018)

2. Soh, J. (2016, March 30) What is IEC 61850 and Why is it Necessary? Phoenix Contact, South East Asia-Blog https://blog.phoenixcontact.com/marketing-sea/2016/03/what-is-iec-61850-and-why-is-it-necessary/ (Accessed September 23, 2018)

3. Harris Williams &Co. (Summer 2010) Transmission & Distribution Infrastructure, a Harris Williams & Co. White Paper https://www.harriswilliams.com/sites/default/files/industry_reports/final%20TD.pdf (Accessed September 23, 2018)

4. Amelsang, S. and Appunn, K. (2018, January 5) The causes and effects of negative power prices, Clean Energy Wire CLEW https://www.cleanenergywire.org/factsheets/why-power-prices-turn-negative (Accessed September 23, 2018)

Chapter 7

1. The Environmental Literacy Council (2015), Petroleum History https://enviroliteracy.org/energy/fossil-fuels/petroleum-history/ (Accessed September 23, 2018)

Chapter 8

1. Muller, E. (2013, April 12) China Must Exploit Its Shale Gas, The New York Times https://www.nytimes.com/2013/04/13/opinion/china-must-exploit-its-shale-gas.html?_r=0 (Accessed September 23, 2018)

2. Wong, E. (2013, April 1) Air Pollution Linked to 1.2 Million Premature Deaths in China, The New York Times https://www.nytimes.com/2013/04/02/world/asia/air-pollution-linked-to-1-2-million-deaths-in-china.html (Accessed September 23, 2018)

Chapter 9

1. Gwynne, P. (1975, April 28) The Cooling World, Newsweek http://www.denisdutton.com/newsweek_coolingworld.pdf (Accessed September 23, 2018)

2. Struck, D. (2014, January 10) How the "Global Cooling" Story Came to Be, Scientific American https://www.scientificamerican.com/article/how-the-global-cooling-story-came-to-be/ (Accessed September 23, 2018)

3. Gwynne, P. (2014, May 21) My 1975 Cooling World Story Doesn't Make Today's Climate Scientists Wrong, Inside Science https://www.insidescience.org/news/my-1975-cooling-world-story-doesnt-make-todays-climate-scientists-wrong (Accessed September 23, 2018)

4. Handwerk, B. (2010, May 7) Whatever happened to the Ozone Hole? National Geographic https://news.nationalgeographic.com/news/2010/05/100505-science-environment-ozone-hole-25-years/ (Accessed September 23, 2018)

Chapter 12

1. Carr, F. (2018, May 13) How I became a Talking Head, Trumplandia https://trumplandiamagazine.com/how-i-became-a-talking-head-403867a7b15d (Accessed September 23, 2018)

Chapter 13

1. Biello, D. (2013, January 23) How Much Will Tar Sands Oil Add to Global Warming? Scientific American https://www.scientificamerican.com/article/tar-sands-and-keystone-xl-pipeline-impact-on-global-warming/ (Accessed September 23, 2018)

2. Tait, C. (2012, July 10, updated 2018, April 30) Enbridge slammed for 'Keystone Kops' response to Michigan spill, The Globe and Mail https://www.theglobeandmail.com/report-on-business/industry-news/energy-and-resources/enbridge-slammed-for-keystone-kops-response-to-michigan-spill/article4402752/ (Accessed September 23, 2018)

3. Buffett, W. (2014, February 28) 2013 Annual Report, Berkshire Hathaway http://www.berkshirehathaway.com/2013ar/2013ar.pdf (Accessed September 23, 2018)

Chapter 17

1. California ISO Glossary http://www.energy.ca.gov/glossary/ISO_GLOSSARY.PDF (Accessed September 23, 2018)

REFERENCES

Chapter 18

1. Mullins, J. (2010, September 8) The eight failures that caused the Gulf oil spill https://www.newscientist.com/article/dn19425-the-eight-failures-that-caused-the-gulf-oil-spill/ (Accessed September 23, 2018)

2. Wikipedia (2018, August 30) Exxon Valdez oil spill https://en.wikipedia.org/wiki/Exxon_Valdez_oil_spill (Accessed September 23, 2018)

3. Blinch, R. (2012, July 10) Enbridge handled oil spill like 'Keystone Kops'-NTSB, Reuters https://www.reuters.com/article/usa-enbridge-spill-idUSL2E8IAB3O20120710 (Accessed September 23, 2018)

4. ExxonMobil corporate website, The Valdez oil spill http://corporate.exxonmobil.com/en/environment/emergency-preparedness/spill-prevention-and-response/valdez-oil-spill (Accessed September 23, 2018)

5. Volcovici, V. (2016, July 20) U.S., Enbridge reach $177 million pipeline spill settlement, Reuters https://www.reuters.com/article/us-enbridge-inc-michigan-oilspill-report/u-s-enbridge-reach-177-million-pipeline-spill-settlement-idUSKCN1001S4 (Accessed September 23, 2018)

6. Bomey, N. (2016, July 14) BP's Deepwater Horizon costs total $62B, USA Today https://www.usatoday.com/story/money/2016/07/14/bp-deepwater-horizon-costs/87087056/ (Accessed September 23, 2018)

Chapter 21

1. Islam, S.N. and Winkel, J. (2017, October) Climate Change and Social Inequality*, United Nations Department of Economic & Social Affairs http://www.un.org/esa/desa/papers/2017/wp152_2017.pdf (Accessed September 23, 2018)

2. Maughan, T. (2015, April 2) The dystopian lake filled by the world's tech lust, BBC Future http://www.bbc.com/future/story/20150402-the-worst-place-on-earth (Accessed September 23, 2018)

Chapter 24

1. Union of Concerned Scientists, Environmental Impacts of Natural Gas https://www.ucsusa.org/clean-energy/coal-and-other-fossil-fuels/environmental-impacts-of-natural-gas#.Wyh2vBJKit8 (Accessed September 23, 2018)

About the Author

Terry Etam is a twenty-five-year veteran of Canada's energy business. He has worked at a number of occupations spanning the finance, accounting, communications, and trading aspects of energy, and has written for several years on his own website Public Energy Number One and the widely-read industry site the BOE Report. Mr. Etam has been called an industry thought leader and the most influential voice in the oil patch. He lives in Calgary, Alberta with his family and, for some reason, a little dog.

Mr. Etam can be reached at tetam462@gmail.com.

CPSIA information can be obtained
at www.ICGtesting.com
Printed in the USA
BVHW051031261122
652665BV00009B/511/J